Local Environmental Politics in China

Knowledge and insight in national environmental governance in China is widespread. However, increasingly it has been acknowledged that the major problems in guiding the Chinese economy and society towards sustainability are to be found at the local level. This book illuminates the fast-changing dynamics of local environmental politics in China, a topic only marginally addressed in the literature. In the course of building up an institutional framework for environmental governance over the last decade, local actors have generated a variety of policy innovations and experiments. In large measure these are creative responses to two main challenges associated with translating national environmental policies into local realities. The first such challenge is a 'policy implementation gap' stemming from the absence of the state capacity necessary to the implementation of environmental measures. The second challenge refers to the need for local non-state actors to engage in environmental management; oftentimes such a 'participation gap' contributes to implementation failures. In recent years, we have seen a multitude of initiatives within China at the provincial level and below designed to bridge both 'gaps'. Hence, the central aim of this book is to assess these experiments and innovations in local environmental politics.

This book was originally published as a special issue of the *Journal of Environmental Policy and Planning*.

Genia Kostka is an Assistant Professor in Chinese Business Studies at the Frankfurt School of Finance and Management, Germany. Her main research interests are in local environmental politics and policy making in China. She has conducted extensive fieldwork on local government organization and environmental governance in both rural and urban China, focusing on institutional change and policy innovations at the township, county and municipal levels.

Arthur P.J. Mol is chair and Professor in environmental policy at Wageningen University, the Netherlands, Director of the Wageningen School of Social Sciences, and Professor in environmental policy at Renmin University, Beijing. He has worked extensively in China, among others on water management, industrial risk management, nuclear and renewable energy, participative environmental governance, and green growth. He is also joint editor of *Environmental Politics*.

Local Environmental Politics in China

Challenges and Innovations

Edited by

Genia Kostka & Arthur P.J. Mol

Routledge
Taylor & Francis Group
LONDON AND NEW YORK

First published 2014
by Routledge
4 Park Square, Milton Park, Abingdon, Oxon OX14 4RN
605 Third Avenue, New York, NY 10017

Routledge is an imprint of the Taylor & Francis Group, an informa business

British Library Cataloguing in Publication Data
A catalogue record for this book is available from the British Library

ISBN13: 978-0-415-72329-9 (hbk)

Typeset in Times New Roman
by Taylor & Francis Books

Publisher's Note
The publisher accepts responsibility for any inconsistencies that may have arisen during the conversion of this book from journal articles to book chapters, namely the possible inclusion of journal terminology.

Contents

Citation Information

The chapters in this book were originally published in the *Journal of Environmental Policy and Planning*, volume 15, issue 1 (February 2013). When citing this material, please use the original page numbering for each article, as follows:

Chapter 1
Implementation and Participation in China's Local Environmental Politics: Challenges and Innovations
Genia Kostka & Arthur P.J. Mol
Journal of Environmental Policy and Planning, volume 15, issue 1 (February 2013) pp. 3-16

Chapter 2
Perverse Incentive Structure and Policy Implementation Gap in China's Local Environmental Politics
Ran Ran
Journal of Environmental Policy and Planning, volume 15, issue 1 (February 2013) pp. 17-40

Chapter 3
Environmental Protection Bureau Leadership at the Provincial Level in China: Examining Diverging Career Backgrounds and Appointment Patterns
Genia Kostka
Journal of Environmental Policy and Planning, volume 15, issue 1 (February 2013) pp. 41-64

Chapter 4
Political Modernization in China's Forest Governance? Payment Schemes for Forest Ecological Services in Liaoning
Dan Liang & Arthur P.J. Mol
Journal of Environmental Policy and Planning, volume 15, issue 1 (February 2013) pp. 65-88

Chapter 5
Environmental Activism in Provincial China
Fengshi Wu

Journal of Environmental Policy and Planning, volume 15, issue 1 (February 2013) pp. 89-108

Chapter 6

The Politics of Waste Incineration in Beijing: The Limits of a Top-Down Approach?
Thomas Johnson
Journal of Environmental Policy and Planning, volume 15, issue 1 (February 2013) pp. 109-128

Please direct any queries you may have about the citations to clsuk.permissions@cengage.com

Notes on Contributors

Thomas Johnson, Department of Public Policy, City University of Hong Kong, Tat Chee, Hong Kong

Genia Kostka, Department of Economics, Frankfurt School of Finance and Management, Frankfurt, Germany

Dan Liang, China National Forestry Economics and Development Research Center, State Forestry Administration, Beijing, China

Arthur P.J. Mol, Department of Social Sciences, Wageningen University, Wageningen, The Netherlands

Ran Ran, Department of Political Science, School of International Studies, Renmin University of China, Beijing, China

Fengshi Wu, Department of Government and Public Administration, Chinese University of Hong Kong, Hong Kong

Implementation and Participation in China's Local Environmental Politics: Challenges and Innovations

GENIA KOSTKA* & ARTHUR P.J. MOL**

**Department of Economics, Frankfurt School of Finance and Management, Sonnemannstraße 9-11, 60314 Frankfurt, Germany*
***Department of Social Sciences, Wageningen University, Hollandseweg 1, 6706 KN Wageningen, The Netherlands*

ABSTRACT *With its recently published 12th Five-Year Plan (FYP, 2011–2015) China's leaders have set ambitious national environmental targets and goals for developing a more sustainable economy and society. Past records, however, show that ambitious goals and regulations too often fail due to shortcomings in local implementation and civil society participation. At the sub-national level, economic, political, and social interests continue to dictate the political agenda and the participation of non-state actors remains limited. This article analyses these implementation and participation gaps and reviews recent innovations and experiments to address these gaps in local environmental politics in China. Although many ongoing experiments and new institutional arrangements can be identified, these projects and initiatives remain limited in scope and geographical spread. Further advances in policy enforcement and in opening up policy design to citizens and other non-state actors at the local level are needed in order to turn the article ambitions of the 12th FYP into reality.*

Introduction

China's 12th Five-Year Plan (FYP), passed by the National People's Congress in 2011, constitutes the main economic, social, and environmental blueprint for the period 2011–2015 and has been described as the 'greenest' so far (Fulton, 2011). The 12th FYP includes specific national targets for 2015 for carbon intensity, energy intensity, water consumption, emissions of pollutants, forest coverage rate, and non-fossil fuel energy production and use (Table 1). With this new FYP, China has demonstrated its commitment to sustainable development and formed an answer to the challenges of growing domestic resource scarcity and

Table 1. Key targets of the 12th FYP 2011–2015

Categories	Targets
Economic targets	• Average annual GDP growth 7% (expected) • More than 45 million jobs to be created in urban areas (expected) • Urban registered unemployment below 5% • Prices to be kept generally stable
Economic restructuring	• Rise in domestic consumption • Breakthrough in emerging strategic industries, which should account for more than 8% of GDP • Increase of service sector value-added output by 4%, accounting for 47% of GDP (expected) • Urbanization rate to increase by 4% points to reach 51.5% (expected)
Innovation	• R&D expenditure to account for 2.2% of GDP (expected) • 3.3 patents to be registered per 10,000 people (expected)
Environment and clean energy	• Non-fossil fuel to account for 11.4% of primary energy consumption (binding) • Water consumption per unit of value-added industrial output to be reduced by 30% (binding); • Energy consumption per unit of GDP to be reduced by 16% (binding) • CO^2 emissions per unit of GDP to be reduced by 17% (binding) • Forest coverage rate to rise to 21.66% and forest stock to increase by 600 million m^3 (binding) • Major pollutants COD and SO^2 to be reduced by 8% and reduction of ammonia nitrogen and nitrous oxides by 10% (binding)
Agriculture	• Annual grain production capacity to be no less than 540 million tonnes • Farmland reserves to be no less than 1.818 billion mu (121.2 million ha) (binding)
Livelihood	• Population size less than 1.39 billion (binding) • Increase urban per capita net income and rural per capita income by 7% compared with 2010 (expected) • Life expectancy increased by 1–74.5 years (expected) • Pension schemes to cover all rural residents and 357 million urban residents (binding) • Construction and renovation of 36 million apartments for low-income families • Minimum wage standard to increase annually by no less than 13% on average • Increase rate of 9-year compulsory school enrolment to 93% (restricted) and increase rate of high school enrolment to 87% (expected)
Social management	• Improved public services for both urban and rural residents • Improved democracy and legal system • Better social management system for greater social harmony • More than 10% of all residents will be registered as community volunteers
Reform	• Encourage qualified enterprises to get listed in stock markets • In-depth reform in monopoly industries for easier market entry and more competition • Improved government efficiency and credibility

Notes: Targets are set at the national level and then passed down to provincial governments. Provincial governments, in turn, include targets in the provincial-level planning documents and allocate targets across departments, municipalities and enterprises. Some targets are 'binding' or restricted (*yueshuxing*) targets and others are 'expected' (*yuqixing*) targets. Fulfilling binding targets is nominally required for a cadre to gain promotion and receive bonus payments.

constraints, and related resource-dependent conflicts. The clear message in the 12th FYP is that the leaders recognize that environmental constraints and energy issues are endangering both future economic growth and social harmony in China. Moreover, China's increasing integration in the world economy and its ambition to play a larger role in world politics further motivates and drives the domestic sustainability agenda.

These developments are also reflected in international studies and reports on China's commitment to environmental protection, resource conservation, and sustainability. Earlier analyses that simply criticize China for its one-dimensional focus on economic growth and neglect of environmental protection have been replaced by more balanced studies. China's record over the past decade with respect to sustainability can be summarized as follows: (i) China has rapidly increased its investments in and commitment to environmental sustainability; (ii) it has accelerated the construction of environmental institutions, laws, policies, and organizations; (iii) resource efficiencies and pollution levels per GDP output have improved considerably; but (iv) absolute resource consumption and pollutions levels often increase and only incidentally decrease; and (v) in addition to strong incentives placed on local officials for achieving economic growth, poor local implementation and participation of non-state actors are among the most important factors that explain shortcomings in sustainability improvements. This commitment to environmental protection is reflected in the ambitions of the 12th FYP. Achieving the detailed and rather specific environmental and clean energy targets of the Plan will have to come through sectoral plans and later regulations and guidance. In launching the FYP at the National Peoples' Congress in March 2011 Premier Wen (2011) stated that China would put in place 'well-equipped statistical and monitoring systems for greenhouse gas emissions, energy conservation and emissions reductions' to ensure these policies are tracked and properly implemented. But neither the Plan nor Premier Wen's Work Report paid much attention to the implementation and participation gap.

This article focuses on the shortcomings in policy implementation and participation that play such an important role in determining China's ability to translate the stated national green goals in the 12th FYP into actual outcomes at the local level. Besides highlighting the challenges of implementation and participation in China's local environmental politics, the article identifies, analyses and assesses recent policy experiments and innovations at the local level. In building up its institutional framework for environmental regulation and governance over the last decade, China has engaged in innovation and experimentation (Carter & Mol, 2006; Mol, 2009a). Many of these innovations address two main challenges. The first challenge relates to local state capacity to successfully formulate and implement environmental policy—referred to hereafter as the 'policy implementation gap' (Challenging the Implementation Gap). The second challenge refers to the participation of local non-state actors to improve environmental management, which we call the 'participation gap' (Challenging the Participation Gap). With respect to addressing both 'gaps' we see a multitude of initiatives and experiments within China at the provincial level and below. Hence, the central aim of this article is to identify the diversity of these experiments and innovations in local environmental politics, and to assess to what extent they contribute to closing the implementation and participation gaps. As such this article forms the introduction for the other articles in this special issue, which each reveal specific aspects of these implementation and participation gaps and the advances that China has made to close these gaps.

Challenging the Implementation Gap

The idea of an implementation gap is not restricted to local environmental policies in China, but is rather common throughout the international literature on environmental politics and policy. Divergence between the central government's national environmental policies and the actual outcomes at the local levels makes up the environmental 'policy implementation gap'. Such gaps can—and have been—identified in many national environmental policy systems (Jänicke, 1986; Knill & Lenschow, 2000; Scheberle, 2004). Yet, the causes and characteristics of such environmental policy implementation gaps often vary across different political systems and governance styles.

The existing literature has uncovered a variety of factors that contribute to the green implementation gap in China. First, while national ministries set the overall direction and long-term goals, these aspirations are often diluted as they pass through China's fragmented vertical and horizontal governance structure (Lieberthal & Oksenberg, 1988). A second view explains the gap between national policies and local practices with reference to China's decentralized governing structure, which allows local officials to be choosy about which national policies to faithfully implement and which ones to ignore (Economy, 2004; O'Brien & Li, 1999). Others reject the decentralization premise and argue that the centre still wields substantial power; here formal constraints imposed by the central government are blamed for shortcomings in environmental policy implementation (Ran, 2013).

Another branch of literature points to the frequent divergence of national and local interests, as local leaders tend to place greater emphasis on economic development than on environmental concerns (Kostka & Hobbs, 2012, 2013; Mol & Carter, 2006; Van Rooij, 2006). A conflict between the central policies and local interests exists because national policy-makers at times fail to take into account policies' potential negative impacts on businesses, employment, and taxation revenues. National policies without sufficient local support and legitimacy are only implemented strictly when there is direct and constant attention from the centre. In order to bridge the gap between national priorities and local interests, local government officials often create informal incentives to fulfill environmental objectives by 'bundling' different interests and policies to build support for environmental policies that are sometimes unpopular with local enterprises and the public (Kostka & Hobbs, 2012). Policy implementation outcomes therefore directly or indirectly relate to the different priorities shaping national and local level incentives.

These four approaches focus on how governance practices and decision-making structures shape implementation outcomes, and offer different accounts of why the central government's environmental regulations and policy mandates are often implemented selectively at the local level. On the whole, the literature suggests that when designing implementation strategies to fulfil national environmental mandates, local government officials will tailor their implementation approach to balance and align these actions with other competing local interests and policy mandates. It is thus vital to understand environmental protection as one of a number of competing local priorities, some of which may be in line with environmental protection and some of which may be in conflict. National environmental targets that imply a potentially negative impact on local economic growth or labour markets will not be prioritized. For instance, although the 12th FYP targets an annual average GDP growth of 7%, only 5 of the 31 provinces have set growth rate targets below 10% in provincial FYPs (*China Daily*, 2011), indicating that provinces continue to place most emphasis on growth-focused development. Many provinces in western and central China still want to catch up with more advanced coastal provinces and only prioritize national environmental targets if these are in line with their provincial economic growth aspirations. In Inner Mongolia, for instance, the combination of low levels of economic development and new inflows of resource-focused investments has meant that provincial leaders only half-heartedly implemented energy saving policies during the 11th FYP.

Yet, over the last decade, we see many new policy efforts at the national and local levels to strengthen the channels for environmental enforcement in China. At the national level, the central government has very noticeably deployed more and more administrative measures and tools to enhance compliance with national environmental rules. The most important of these are mandatory 'binding' (*yueshuxing*) environmental targets for government departments and enterprises that include the energy efficiency and carbon emission targets in the 12th FYP. These binding environmental targets are written into local cadres' annual responsibility contracts and are crucial criteria in cadre promotion decisions, thereby incentivizing officials at each layer of government administration to fulfil upper level governments' environmental mandates. The introduction of specific binding targets not only indicates that environmental protection is a national priority, but also serves as an important signal to local governments in how they should prioritize competing interests and priorities at the local level. As such, targets in the cadre responsibility contracts have an important steering function within China's decentralized economy and are increasingly used by national leaders to ensure local leaders' commitment to environmental protection, resource conservation, and sustainability.

Moreover, the central government is continuously improving the design and usage of administrative tools and measures. The setting of environmental targets for particular sectors and regions has become increasingly sophisticated. Improvements include expansion of binding targets from energy efficiency to many other environmental areas, use of tailored target setting methods rather than a 'one-size-fits-all' approach, increased emphasis on annual targets over aggregated 5-year targets, and increased guidance on target measurement and verification standards to local implementing agencies. Beijing has also incorporated more bottom-up feedback elements in the planning process in order to set more realistic goals and secure provincial leaders' support. For instance, provincial Development and Reform Commissions and the Economic Commissions were asked to comment on the new national energy target in the 12th FYP before targets were officially announced in 2011 (Ohshita & Price, 2011). In addition, China secured international support to help meet its environmental, energy- and carbon-intensity targets. In 2012, for instance, China signed a €25 million finance agreement with the European Commission to obtain technical assistance, training and know-how in areas such as water, waste and heavy metal pollution, emission trading systems, and sustainable urbanization (European Commission, 2012). Moreover, local environmental protection bureaus (EPBs) increasingly rely on environmental courts to meet their environmental pollution targets and fulfil national standards. Despite numerous shortcomings in the current environmental litigation system (Stern, 2011), environmental courts are growing fast across China and by 2012, nearly 100 environmental courts had been opened in 16 provinces (Zhang & Zhang, 2012). While such improvements in administrative tools and environmental compliance do not guarantee better local policy implementation, they suggest a shift in government officials' mindsets and they lay the institutional foundation for a more robust environmental protection regime.

China is also experimenting with a variety of market-based instruments to supplement existing administrative tools. In 2009, Beijing, Shanghai, and Tianjin created the first voluntary carbon cap-and-trade schemes, and by 2010, a total of 12 voluntary climate change exchanges had come into being (Fulton, 2011). Currently, these voluntary exchanges serve as platforms for small- or medium-scale transactions, but many grey areas persist with regards to monitoring, reporting, and verification of emissions data. China is also taking its first step towards a market-based electricity pricing system that more accurately reflects the actual costs of power generation. For industrial users, electricity prices have been repeatedly raised over the last years, with the most recent price raise in June 2011 for 15 provinces. For residential users, a three-tiered new electricity pricing system was rolled out in July 2012, in the hope that the price increases for the higher two tiers would encourage consumers to save electricity (*China Daily*, 2012). Market-based instruments are also increasingly mixed with command and control mechanisms to improve water, air, and soil management in China. For instance, economic instruments have already been widely introduced in the water and wastewater sector, with notable results (Zhong & Mol, 2010). Moreover, payment for environmental services (PES) is soaring in China, but is often a mixture of government-directed PES schemes with some market elements (Liang & Mol, 2013).

In addition to using a variety of administrative and market-based tools, the creation of new decision-making structures has also improved local implementation of environmental issues. It is not uncommon to have more than 10 different government bureaus and agencies involved in the implementation of a single policy. To overcome implementation problems resulting from this fragmented bureaucratic structure, local governments are increasingly using inter-departmental *ad hoc* coordination committees, known as leading small groups (*lingdao xiaozu*). Previously, such *ad hoc* coordination mechanisms were (and are) used at the national level to build consensus across government agencies (Zhu, 2010). For instance, well-known environmental working groups at the national level include the National Coordination Committee on Climate Change and the State Council Coordination Committee for Energy Conservation and Emission Reduction. Over the last few years, provincial, municipal, and county governments have also set up their own coordination committees to develop environmental protection strategies across departments and bureaus. Such local inter-departmental working groups are usually headed by a local mayor or vice mayor, and in the event of disagreement between different departments, it falls to the leader of a working group to decide the matter.

Environmental implementation also improves as provinces and cities increasingly take on a leading role in transitioning to a greener growth path. Many local governments are developing their own strategies for environmental protection and an implicit competition has emerged between cities and provinces claiming to be at the forefront of environmental innovation. Some local governments have even introduced local environmental regulations that are more stringent than those of the centre, such as Kunming's water pollution policy (Van Rooij, 2006). The typical Chinese policy of working with 'model' cities or companies and awarding prizes has contributed to such competition.[1] Previously, examples included 'State Environmental Protection Model Cities' such as Dalian and Xiamen. Increasingly, traditionally pro-growth cities such as Guangzhou, Wuhan, and Chengdu are becoming more environmentally friendly (Lo *et al.*, 2006). But even small municipal-level cities are entering this implicit competition between local governments. For example, Dezhou in Shandong has become China's self-styled 'solar valley' by integrating solar power into its street lighting, heating and transport systems.

Challenging the Participation Gap

In many countries around the world, implementation failures in environmental policies have resulted in increasing initiatives to enlarge the participation of non-state actors in both policy formulation and policy implementation. It is believed that participation can help overcome some of the implementation shortcomings of conventional state-led and hierarchical policy models. Among others, participation brings in more information and knowledge on activities to be regulated, builds commitment at an early stage among those confronted with policies and measures that aim to restrict or condition behaviour, and reduces transaction costs later on in the policy process. Regardless of the many claims regarding the environmental gains of more participatory policy models and measures, empirical evidence, indicating that more participation provides better environmental outcomes is still rather thin (Bulkeley & Mol, 2003; Newig & Fritsch, 2009). But participation is considered to have a number of other benefits to environmental policy (Beierle, 1999): it educates the public; it incorporates public values, assumptions, and preferences into decision making; it increases the substantive quality of decisions; it fosters trust in institutions; it reduces conflict; and it makes decisions more cost-effective. Finally, participation in environmental policy-making and implementation also has a more normative background, as it is strongly rooted in ideas of democratic governance. Through participatory methods, citizens and interest groups have better access to and influence over policy processes, thus enlarging the legitimacy and democratic character of these policies.

China's environmental governance systems and initiatives rely predominantly on administrative measures that are driven, initiated, and executed by the government. Participation in China's environmental policy-making has traditionally been structured and institutionalized within the state and party systems, for instance, through the China People's Political Consultative Conference at various levels. In these institutions, different local, sectoral, and thematic interests could participate in a formalized and structured process of representation and consultation. These mechanisms of participation—characterized by participatory possibilities that are rather restricted and with access limited to a few individuals—continue to exert influence over Chinese (environmental) policy-making. However, over the last decade and especially in the field of environmental politics, we can witness the flourishing of a variety of experiments and new institutional arrangements, which allow for further participation in China's policy-making and implementation, at various levels of governance.[2] Environmental policies seem to have become the leading field of such participatory experiments, enabled by the acknowledgement among Chinese leaders of the need for more counter-vailing environmental participation to check the growing influence of narrow local government preferences or even private economic interests and powers.

The number of environmental non-governmental organizations (ENGOs) has been rapidly increasing in China, including both officially registered and unregistered groups. While reliable official numbers do not exist, estimates suggest that there are approximately 1000 registered ENGOs, as well as a similar number of unregistered ENGOs. The majority of these ENGOs are quite small and are not directly engaged in environmental policy-making and implementation, but rather focus on awareness raising, education, study, and research. A number of those engaged in policy advocacy are closely related to governmental organizations and institutes and are often referred to as government organized NGOs (GONGOs). Through closed networks with policy-makers and their expert knowledge, these GONGOs articulate environmental interests and bring them into state institutions and decision-making processes. In doing so, GONGOs play a role in bridging the gap between NGOs and civil society on the one hand, and the state on the other, thus 'becoming an important, non-state arena for China's environmental politics' (Wu, 2002, p. 48). But increasingly, we see also independent ENGOs that participate in environmental policy-making and implementation processes, by bringing in knowledge, writing petitions, using media outlets, discussing alternatives with government officials, providing legal assistance for pollution victims to make them aware of their rights, and higher levels of what Johnson (2010) calls rule-based activism (such as performed by the famous NGO 'Center for Legal Assistance to Pollution Victims'). Several studies have detailed how such ENGOs explore the boundaries of what is allowed in contemporary China in terms of NGO engagement, policy involvement and protest (cf. Hildebrandt, 2011; Wu, 2013; Xie, 2007).

Public hearings form a more institutionalized arrangement for participation in China's environmental policy-making. The best-known example concerns public hearings in the Environmental Impact Assessment (EIA) procedure, as formalized in the 2002 EIA law and its implementation measures. Zhao's (2010) study of a public hearing concerning renovations to the Imperial Summer Palace (Yuan Ming Yuan Park) and Johnson's (2013) study on the siting of waste incinerators illustrate that such a legal obligation to organize consultation with citizens does not mean that it always takes place in a meaningful way. Another example of participation through public hearings takes place in the field of setting tariffs for drinking water and wastewater (Zhong & Mol, 2008). Here, various interest groups as well as citizens participate in a public hearing on plans to increase drinking water tariffs and wastewater fees. Any government proposal needs to have at least two-third majority of the votes in the public hearings in order to be able to be implemented.

By the early 1990s, China had already set up systems of complaints (the so-called letters and visits) to assist government monitoring and priority setting systems in the field of environmental pollution. The public participation in these systems, and hence the number of visits and letters addressed to governmental authorities, has grown over the years. More recently, hotlines and digital complaint systems, and experiments with so-called 'green bounty hunter' systems[3] have further enhanced public participation in environmental enforcement of governmental agencies (Bretell, 2007; Mol, 2010; Wu, 2013; Zhang, 2011). There are clear indications that these complaint systems do have an impact on priority setting and control and enforcement activities of local environmental officials.

Participation is also enhanced through information disclosure, for quite some time, a sensitive issue in China. China joined the global trend of environmental information disclosure regulations (cf. Mol, 2006) rather late, with the entering into force in 2008 of the Open Governmental Information Regulation (OGIR) and the related Environmental Information Disclosure Decree (EIDD). The main motivations behind OGIR are to alleviate the information asymmetry between actors; to gain political credits by 'improving the transparency of governmental work'; and to improve administrative performance (Wang *et al.*, 2008). The EIDD is in line with a wider system of information disclosure and public access to environmental information (such as in the Cleaner Production Promotion Law and the Green Watch programme; Li & Xiong 2005; Mol, 2009b). The Decree requires not only environmental authorities, but also industries to disclose environmental information, and specifies how and in what time limits, this should be done. In addition, the Decree requires the establishment of monitoring, evaluation, and supervision systems. Those organizations violating the rules of information disclosure will be held responsible and accountable by their higher level supervisors; and industries violating the rules will face penalties. Citizens can sue public authorities and industries that violate the Decree. But state secrets, commercial secrets, individual privacy, national security, public security, economic security, and social stability are all considered legitimate reasons to withhold disclosure of (environmental) information. Many of the terms in the Decree are also not clearly defined, making it easy for companies and officials to withhold information. Evaluation of the EIDD implementation after 2 years showed that it failed to trigger major participation of citizens, which was also due to poor implementation and functioning of the information disclosure systems of provinces and companies (Greenpeace China, 2009; He *et al.*, 2011; IPE & NRDC, 2010; Mol *et al.*, 2011; Zhang *et al.*, 2010).

It is also interesting to note the degree of participation in local PES programmes, for instance, in the protection of forests (Liang & Mol, 2013). After the serious flooding in the late 1990s, the Chinese government has enlarged the area of so-called public benefit forest, and installed a considerable number of payment schemes to compensate forest owners for the restrictions in use of these forests. Forest right-holders (farmers, collective owners such as villages, and governmental forest services) have been involved in the demarcation of the public benefit forest falling under these schemes, the design of forest management contracts, and the examination and enforcement of implementation, but have hardly been involved in the formulation of these PES policies. Such participation did improve legitimacy, acceptance, and implementation of these policies for enlarging public benefit forest. But, as Bennett (2009, p. 72) concludes, 'significant potential exists for greater private sector participation in and funding of PES/MES initiatives, and in particular regarding watershed ecosystem services in China's wealthier regions'.

Regardless of these experiments and new institutional arrangements, the practice of participation of concerned citizens, of farmers and of NGOs in designing projects and policies that are of consequence to the environment remains limited, for several reasons. First, most of these experiments relate to policies of the Ministry of Environmental Protection or its subsidiaries at lower levels. New participatory arrangements hardly touch upon policies and projects of the more influential ministries, such as those of National Development and Reform Commission (NDRC) or of the industrial ministries. Second, most of the participatory experiments relate to implementation practices and are helpful in strengthening the implementation, control and enforcement of existing regulations, plans, policies, and projects. New forms of environmental participation at the level of designing policies and practices are much more rare. Research into the law-making process of the 2002 Cleaner Production Promotion Law (Mol & Liu, 2005) showed that various governmental bodies had quite some influence in changing the law between its initial draft and the final accepted law, but among them were no environmental NGOs or concerned citizens outside the formal governmental or legal expert circles. Third, with respect to participation in policy implementation, environmentalists are highly dependent on local conditions. The conclusion emerges that national leaders increasingly value environmental participation as they see it as a necessary counter-vailing power against the forces of economic market actors, whether purely privatized versions or state-related market interests. However, this is not the case in many localities, especially those outside the more visual and internationally connected ones (such as the main economic centres: Beijing, Shanghai, and Guangdong). In these 'peripheral' localities, local government and business leaders continue making decision behind closed doors and strongly suppress attempts to participate in environmental decision-making and implementation, either by concerned citizens, NGOs or the local media. Here, we still see many of the conventional Chinese reflexes towards democratic and participatory sprouts of environmental governance. Fourth, recent research on collective activism in rural China shows that rural communities sometimes fall into a 'compensation trap', where reliance on compensation from factories prevents action against the source of pollution (Van Rooij *et al.*, 2012). This means that famers end up relying on, and therefore, not opposing, pollution. Reliance on compensation becomes then another obstacle to public participation.

In summary, the impact of China's national commitments to increase environmental protection will largely depend on how effectively environmental laws and regulations are enforced at the local level. While national advancements and ambitious goal-setting are promising, environmental policies are still widely resented at the sub-national level because local interests often diverge from national priorities and central directives are also partially in conflict with each other. Recent efforts and experiments to close this implementation gap include improvements in environmental administrative tools and planning processes, experiments with market-based instruments, and new governing decision-making structures. And the newly emerging institutional arrangements that allow for further participation in China's environmental policy-making, such as increased environmental NGO activism, public hearings, and improved information disclosure, also contribute to enhance local environmental policy implementation, and sometimes even policy-making.

The Contributions to this Special Issue

The above analysis of the implementation and participation gaps in local environmental politics suggests that there is still much to discover about this issue, especially regarding the mechanisms, effects, and impacts of on-going experiments and innovations to close these gaps. Five contributions in this special issue on local environmental politics analyse recent innovations at the local level by drawing on examples from waste management, pollution, and forest management. The articles study environmental policies in eight provinces across China: Beijing, Fujian, Guangdong, Guangxi, Inner Mongolia, Liaoning, Shanxi, and Xinjiang. The first contribution analyses central–local relations (Ran, 2013); the second and third contributions look at implementation agencies and policies (Kostka, 2013; Liang & Mol, 2013), while the fourth and fifth contributions analyse public participation in local environmental politics (Johnson, 2013; Wu, 2013).

The contribution by Ran (2013) evaluates the origin of the environmental implementation gap. Drawing from fieldwork in the four cities in Fujian, Xinjiang, Liaoning, and Jiangsu, Ran challenges the conventional view in the existing literature that environmental deterioration in China is predominantly caused by local governments who turn a blind eye to environmental pollution and degradation. Instead, she argues that in China, the central government provides insufficient incentives to officials in local policy implementation to motivate them to enforce national environmental policies and laws. The article's analysis of existing political, financial, and moral incentive structures helps to explain why many national environmental policies have often produced outcomes at the local level with little concrete effect. By drawing attention to the 'perversely structured incentive system' in China, this article sheds much light on the reasons behind continuing problems in local environmental politics.

The article by Kostka (2013) analyses the career backgrounds and appointment patterns of departmental heads in provincial EPBs. Using biographical information and drawing on fieldwork in Shanxi and Inner Mongolia, the study shows that more than three quarters of China's provincial EPB directors are promoted from outside the EPB structure, partly as a control mechanism but also in order to strengthen EPBs' links to other bureaus. While their non-environmental background enables EPB leaders to link with other key economic and political actors to gain support for environmental programmes, it also implies that decision makers often lack environmental training and experience required to make informed decisions dealing with environmental management. EPB directors' different career backgrounds also shed light on appointment preferences of provincial leaders. While leaders in Shanxi, a province under external pressure to improve air pollution, selected an EPB director with the skills and credentials to effectively implement environmental mandates, leaders in still-developing Inner Mongolia appointed a candidate who could balance economic growth and environmental protection concern.

Liang & Mol's (2013) article critically examines forest policy and payment schemes in two counties in Liaoning province. The study analyses this issue by making use of 54 interviews with farmer household heads on the topic of how the new PES in local forest protection affected farmers' livelihood and participatory opportunities. The main finding of their analysis is that the new schemes made policy implementation more participatory and inclusive. However, although the payment schemes on environmental services increased farmers' participation and compliance, the authors point out that the schemes did not improve farmers' livelihood significantly. These results may have important implications for local payment schemes underway in China and many other parts of the world.

The final two articles further analyse participation of non-state actors. The contribution by Wu (2013) examines ENGO development in Guangdong and

Guangxi provinces. Two factors that help to explain diverging ENGO growth at the provincial level are the development of civil society within the province and existing local government support. Wu's findings highlight the importance of understanding the provincial political context and institutional contingencies when examining local environmental activism.

In the final article of this special issue, Johnson (2013) presents a study of three anti-incinerator campaigns in Beijing. Through documentary analysis and in-depth stakeholder interviews, Johnson vividly illustrates how anti-incineration campaigns have partly undermined the government's top-down, non-consultative approach to waste management. Yet, the three case studies also demonstrate that anti-incinerator campaigns in China are predominantly localized and issue-focused with slim prospects for disparate campaigners to form an alliance strong enough to counterbalance the state's incineration policy.

Conclusion

This special issue provides detailed insights into local environmental politics in China. The articles unpack how Chinese government officials, enterprise managers, farmers, and public activists use existing and creating new institutional spaces for furthering engagement in local environmental politics. The articles also analyse how these actors are engaged in and debate environmental policy formulation and enforcement at the local level in China. The articles highlight the high degree of variation in local environmental politics, with respect to both regions and environmental subject areas. The key contribution of this special issue to the existing literature on environmental policy implementation is the better understanding of the behaviour of citizens, NGOs, and local governments in China's changing local environmental politics. In doing so, the various contributions go beyond highlighting the shortcomings of local implementation by drawing attention to the innovativeness of these local agents in developing solutions to environmental implementation and participation challenges in China, and in creating new institutional spaces for participation and policy implementation.

Notes

1. In August 2010, the NDRC selected eight cities (Baoding, Chongqing, Guiyang, Hangzhou, Nanchang, Shenzhen, Tianjin, and Xiamen) and five provinces (Guangdong, Hubei, Liaoning, Shaanxi, and Yunnan) as low-carbon pilot provinces and cities. Each of the selected pilot regions will develop comprehensive and locally tailored plans to reduce greenhouse gas emissions.
2. Martens (2006) shows that there are also opportunities for Chinese citizen-consumers to participate in environmental management beyond the political domain (e.g. in nature protection, in sustainable household practices, in constructing green company images). However, these participatory methods go beyond the scope of this article.
3. These cash-reward informant programmes provide a financial reward to citizens that report environmental violations. It was first experimented in Fuyang in 2003, and later adopted by many other cities and provinces, with variation in success rate (Zhang, 2011).

References

Beierle, T. C. (1999) Using social goals to evaluate public participation in environmental decisions, *Review of Policy Research*, 16(3–4), pp. 75–103.

Bennett, M.T. (2009) *Markets for Ecosystem Services in China: An Exploration of China's "Eco-Compensation" and Other Market-Based Environmental Policies* (Washington DC: World Bank).

Bretell, A. (2007) China's pollution challenge: The impact of economic growth and environmental complaints on environmental and social outcomes, in: S. Guo & B. Guo (Eds), *Challenges Facing Chinese Political Development*, pp. 155–192 (Lanham: Rowman and Littlefield).

Bulkeley, H. & Mol, A. P. J. (2003) Participation and environmental governance: Consensus, ambivalence and debate, *Environmental Values*, 12(2), pp. 143–154.

Carter, N. T. & Mol, A. P. J. (2006) China and the environment: Domestic and transnational dynamics of a future hegemon, *Environmental Politics*, 15(2), pp. 331–345.

China Daily (2011) China prepares to end GDP obsession, 7 March. Available at http://www.chinadaily.com.cn/bizchina/2011-03/07/content_12129931.htm (accessed 16 June 2011).

China Daily (2012) China implements new electricity pricing system, 2 July. Available at http://www.chinadaily.com.cn/business/2012-07/02/content_15542580.htm (accessed 24 September 2012).

Economy, E. C. (2004) *The River Runs Black: The Environmental Challenge to China's Future* (Ithaca, NY: Cornell University Press).

European Commission (2012) *The European Union and China join forces to address environment, urbanisation and climate change challenges*, 20 September 2012. Available at http://europa.eu/rapid/pressReleasesAction.do?reference=IP/12/989&format=HTML&aged=0&language=EN&guiLanguage=en (accessed 24 September 2012).

Fulton, M. (2011) *12th Five Year Plan—Chinese leadership towards a low carbon economy.* Deutsche Bank Group DB Climate Change Advisors. Available at http://www.dbcca.com/dbcca/EN/_media/China_12th_Five_Year_Plan.pdf (accessed 9 September 2012).

Greenpeace China (2009) *Silent Giants an Investigation into Corporate Environmental Information Disclosure in China* (Beijing: Greenpeace).

He, G.-Z., Zhang, L., Lu, Y.-L. & Mol, A. P. J. (2011) Managing major chemical accidents in China: Towards effective risk information, *Journal of Hazardous Materials*, 187(1–3), pp. 171–181.

Hildebrandt, T. (2011) The political economy of social organization registration in China, *The China Quarterly*, 208, pp. 970–989.

IPE & NRDC (Institute of Public & Environmental Affairs, Natural Resources Defense Council) (2010) *Progress and Retreat for Environmental Transparency in China: Announcing the 2009–10 Pollution Information Transparency Index (PITI) Results* (Washington, DC: NRDC) (in Chinese).

Jänicke, M. (1986) *Staatsversagen. Die Ohnmacht der Politik in die Industriegesellschaft* (München: Piper).

Johnson, T. (2010) Environmentalism and NIMBYism in China: Promoting a rule-based approach to participation, *Environmental Politics*, 19(3), pp. 430–448.

Johnson, T. (2013) The politics of waste incineration in Beijing: The limits of a top-down approach? *Journal of Environmental Policy and Planning*, 15(1): China's Local Environmental Politics, pp. 109–128.

Knill, C. & Lenschow, A. (Eds) (2000) *Implementing EU Environmental Policy: New Directions and Old Problems* (Manchester: MUP).

Kostka, G. (2013) Environmental protection bureau leadership at the provincial level in China: Examining diverging career backgrounds and appointment patterns, *Journal of Environmental Policy and Planning*, 15(1): China's Local Environmental Politics, pp. 41–63.

Kostka, G. & Hobbs, W. (2012) Local energy efficiency policy implementation in China: Bridging the gap between national priorities and local interests, *The China Quarterly*, 211, pp. 765–785.

Kostka, G. & Hobbs, W. (2013) Embedded interests and the managerial local state: The political economy of methanol fuel-switching in China, *Journal of Contemporary China*, 22(80), 15(1): China's Local Environmental Politics, pp. 61–84.

Li, F. G. & Xiong, B. (2005) 环境信息公开及在中国的实践，中国人口资源与环境, 15(4), pp. 22–25 [Public access to environmental information and its practice in China, *Resources and Environment*, 15, 4, pp. 22–25].

Liang, D. & Mol, A. P. J. (2013) Political modernisation in China's forest governance? Payment schemes for forest ecological services in Liaoning, *Journal of Environmental Policy and Planning*, 15(1): China's Local Environmental Politics, pp. 65–88.

Lieberthal, K. & Oksenberg, M. (1988) *Policy-Making in China: Leaders, Structures, and Processes* (Princeton, NJ: Princeton University Press).

Lo, C., Fryxell, G. & Wong, W. (2006) Effective regulations with little effect? The antecedents of the perceptions of environmental officials on enforcement effectiveness in China, *Environmental Management*, 38(3), pp. 388–410.

Martens, S. (2006) Public participation with Chinese characteristics; Citizen consumers in China's environmental management, *Environmental Politics*, 15(2), pp. 63–82.

Mol, A. P. J. (2006) Environmental governance in the information age: The emergence of informational governance, *Environmental Planning C*, 24(4), pp. 497–514.

Mol, A. P. J. (2009a) Urban environmental governance innovations in China, *Current Opinion in Environmental Sustainability*, 1(1), pp. 96–100.

Mol, A. P. J. (2009b) Environmental governance through information: China and Vietnam, *Singapore Journal of Tropical Geography*, 30(1), pp. 114–129.

Mol, A. P. J. (2010) Sustainability as global attractor: The greening of the 2008 Beijing Olympics, *Global Networks*, 10(4), pp. 510–528.

Mol, A. P. J. & Carter, N. T. (2006) China's environmental governance in transition, *Environmental Politics*, 15(2), pp. 149–170.

Mol, A. P. J. & Liu, Y. (2005) Institutionalising cleaner production in China: The cleaner production promotion law, *International Journal of Environment and Sustainable Development*, 4(3), pp. 227–245.

Mol, A. P. J., He, G.-Z. & Zhang, L. (2011) Information disclosure as environmental risk management: developments in China, *Journal of Current Chinese Affairs*, 40(3), pp. 163–192.

Newig, J. & Fritsch, O. (2009) Environmental governance: Participatory, multi-level—and effective? *Environmental Policy and Governance*, 19(3), pp. 197–214.

O'Brien, K. & Li, L. (1999) Selective policy implementation in rural China, *Comparative Politics*, 31(2), pp. 167–186.

Ohshita, S. & Price, L. (2011) March 2011 Update—*Target allocation for China's provinces: Energy intensity in the Five-Year Plan*, Lawrence Berkeley National Laboratory. Available at http://china.lbl.gov/publications/target-allocation-methodology-provinces-china (accessed 9 September 2012).

Ran, R. (2013) Perverse incentive structure and policy implementation gap in China's local environmental politics, *Journal of Environmental Policy and Planning*, 15(1): China's Local Environmental Politics, pp. 17–39.

Scheberle, D. (2004) *Federalism and Environmental Policy. Trist and the Politics of Implementation* (Washington, DC: Georgetown University Press).

Stern, R. (2011) From dispute to decision: Suing polluters in China, *The China Quarterly*, 206, pp. 294–312.

Van Rooij, B. (2006) Implementation of Chinese environmental law: Regular enforcement and political campaigns, *Development and Change*, 37(1), pp. 57–74.

Van Rooij, B., Lora-Wainwright, A., Wu, Y. & Zhang, Y. (2012) The compensation trap: The limits of community-based pollution regulation in China, *Pace Environmental Law Review*, 29(3), pp. 701–745.

Wang, H., Greer, L. & Lin, X. (2008) 环境信息公开的实践及启示，世界环境, 5, pp. 24–26 [Practice and inspiration of environmental information disclosure, *World Environment*, 5, pp. 24–26].

Wen, J. (2011) *Report on the work of the government*. Delivered at the Fourth Session of the Eleventh National People's Congress on 5 March 2011. Available at http://online.wsj.com/public/resources/documents/2011NPCWorkReportEng.pdf (accessed 9 September 2012).

Wu, F. (2002) New partners or old brothers? GONGOs in transitional environmental advocacy in China, *China Environment Series*, 5, pp. 45–58.

Wu, F. (2013) Environmental activism in provincial China: Comparative evidence from Guangdong and Guangxi, *Journal of Environmental Policy and Planning*, 15(1): China's Local Environmental Politics, pp. 89–108.

Xie, L. (2007) *Environmental Activism in China* (London and New York: Routledge).

Zhang, X. (2011) Green bounty hunters: Engaging Chinese citizens in local environmental enforcement, *China Environmental Series*, 2010/2011, pp. 137–153.

Zhang, M. & Zhang, B. (2012) Specialized environmental courts in China: Status quo, challenges and responses, *Journal of Energy & Natural Resources Law*, 30(3), *forthcoming*, August 2012. Available at http://papers.ssrn.com/sol3/papers.cfm?abstract_id=2119866 (accessed 9 September 2012).

Zhang, L., Mol, A. P. J., He, G. & Lu, Y. (2010) An implementation assessment of China's Environmental Information Disclosure Decree, *Journal of Environmental Sciences*, 22(10), pp. 1649–1656.

Zhao, Y. (2010) Public participation in China's EIA regime: Rhetoric or reality? *Journal of Environmental Law*, 22(1), pp. 89–123.

Zhong, L. & Mol, A. P. J. (2008) Participatory environmental governance in China: Public hearings on urban water tariff setting, *Journal of Environmental Management*, 88(4), pp. 899–913.

Zhong, L. & Mol, A. P. J. (2010) Water price reforms in China: Policy-making and Implementation, *Water Resources Management*, 24(2), pp. 377–396.

Zhu, X. (2010) *China's national leading group to address climate change: Mechanism and structure*. EAI Background Brief No. 572. Available at http://www.eai.nus.edu.sg/BB572.pdf (accessed 9 September 2012).

Perverse Incentive Structure and Policy Implementation Gap in China's Local Environmental Politics

RAN RAN

Department of Political Science, School of International Studies, Renmin University of China, Haidian District, Beijing 100872, People's Republic of China

ABSTRACT *The obvious paradox within China's environmental politics is the big gap between the central government's policy and its implementation outcomes at local levels. Despite concerns about implementation at the local level, much about the role of central governments in China's local environmental politics is still poorly understood. This article examines how the incentive structure set by the central authorities affects the policy implementation gap at the local level. Drawing on fieldwork and document analysis, this article argues that the incentives set by the central government regarding environmental policy implementation at local levels are perversely structured, meaning that the central government provides much more incentive for local governments' non-implementation or poor implementation of its environmental policies than it provides for full implementation. The central government's failure to encourage—politically, financially, as well as morally—local government officials to appropriately implement environmental policies can partly explain the production of the policy implementation gap at the local level. This implementation gap cannot be overcome by efforts at the local level unless the central government takes significant measures to address the perverse incentive structure embedded in the overall structure of China's local environmental politics.*

Introduction

The discourse of China's rising, a result of the country's economic growth in the past three decades, does not lead people to neglect the tremendous environmental cost behind its economic miracle. By the end of 2010, China had already become the world's second-largest economy. At the same time, after becoming the world's largest emitter of greenhouse gases in 2008, China passed the USA to become the world's biggest energy consumer in 2010. Few people would doubt that China is one of the most polluted countries in the world. More than 700,000 Chinese die each year as a result of pollution (World Bank & SEPA, 2007). According to the vice-minister of the Ministry of Environmental Protection (MEP) Pan (2006), pollution costs the Chinese economy at least 8–15% of its GDP each year, which does not include the staggering costs associated with health problems and human suffering. The number of complaints received by environmental authorities has increased by 30% each year, while the number of mass protests related to environmental issues has grown by 29% each year since 2002 (Zhou, 2008). Despite the stricter environmental regulations included in the recent 11th and 12th Five Year Plans (FYPs), environmental degradation and the tremendous social and economical problems that have been generated are widely recognized as one of the most pressing challenges for a rising China, domestically as well as globally (Economy, 2007; Mol, 2011; Watts, 2010).

While much research focuses on how economic development, demographic change, international trade and globalization have influenced and will continue to influence the environment, discussion of the role of political factors has also contributed significantly to understanding China's environmental trajectory. This discussion has gradually developed into the environmental politics approach for environmental study in China.[1] Researchers of environmental politics are interested in exploring how China's government responds to domestic environmental, energy, and sustainable development difficulties. They clearly point out that the governing structure, bureaucratic hierarchy, policy-making and implementation process, and regulatory patterns as the factors critical to understanding the challenges of environmental governance in China (Wu, 2009).

Several researchers have noticed a paradox within China's environmental governance: on the one hand, Chinese central government leaders show a great awareness of environmental problems and have provided a high-quality framework for pursuing sustainable development and environmental progress by constructing a comprehensive and modern set of environmental laws. They have also established a formal environmental administration department and participate actively in international environmental protection activities (Economy, 2004; OECD, 2007). The 12th FYP has set stricter environmental and energy targets. On the other hand, however, many environmental policies have produced outcomes with little concrete effect (Greenpeace, 2011). Therefore, whether and to what degree the Chinese authorities' seemingly significant efforts to protect the environment have been translated into environmental outcome in practice have become major questions in the study of Chinese environmental politics (Tang *et al.*, 2005; Tilt, 2007; Van Rooij & Lo Wing-Hung, 2010). One of the well-accepted arguments in the existing literature is that the major obstacle to environmental governance lies at the local level. Elizabeth Economy points out that China's central government has given too much authority to local officials, retaining only weak control over policy implementation. In her view, policies are ineffective in large part because local officials oppose the central government's policies or refuse to implement them properly (Economy, 2004). Van Rooij (2006) argues that loose enforcement is rooted in a lack of local legitimacy caused by conflicts of interest between national regulations and local stakeholders.

Local governments and officials who turn blind eyes to polluters in their jurisdictions are often charged with being the largest contributors to the environmental policies implementation gap. However, in order to better understand the causes of the implementation gap, it's also necessary to highlight the institutional constrains that are embedded in China's local environmental politics. Kenneth Lieberthal attempted to understand the disjuncture between the Chinese government's promise and its performance on environmental issues by analysing two systematic dynamics of China's governing system: (a) the distribution of authority and (b) the structure of incentives. He noted that

> China's present political system operates in a rather fluid fashion, with great local variation, considerable opportunity for local initiative, and tremendous pressure on local officials to give priority to rapid economic development. As a consequence, the country's implementation of its extensive environmental regulations is very inadequate, and its environmental offices generally are under the authority of officials whose priority is short term growth rather than long term sustainability. (Lieberthal, 1997, p. 8)

Enlightened by Lieberthal's top-down approach, this research also tries to bring the central government back into view by looking at how the incentive structure set by the central government shapes the policy implementation gap through its effect on China's local environmental politics?[2] Drawing on empirical findings from fieldwork in four cities, I argue that the incentives set by the central government regarding environmental policy implementation at local levels are perversely structured, meaning that the central government provides much more incentive for local governments' non-implementation or poor implementation of its environmental policies than it provides for full implementation. The central government's failure to encourage—politically, financially, as well as morally—local government officials to appropriately implement environmental policies can partly explain the production of a policy implementation gap in China's local environmental politics.

Conceptual Framework and Data

Organizational behaviour literature has pointed out that, because humans are purposeful creatures, the study of incentive systems is vital for understanding individual and organizational activity (Clark & Wilson, 1961). Public policy scholars differentiate between two types of incentives: financial and non-financial. The financial incentives are largely payment and direct material benefits; non-financial incentives could be promotion, gifts, rewards, and other forms that are not direct financial incentives. The value of non-financial incentives is psychological (Lopes & Theisohn, 2004). Studies on the performance of public service have shown that financial incentives constitute a necessary, but not sufficient, condition for motivating enhanced performance from civil servants (Ulleberg, 2009). Other studies have illustrated the limited impact of financial incentives, especially in the health and education sectors, where conscience, vocation, and professional ethos are important motivational factors (Mathauer & Imhoff, 2006). Environmental economists stress the rising role of incentive-based mechanisms, such as tradable permits and green taxation, in environmental policy regulation (Hahn & Stavins, 1992; Pennington, 1999). Some studies on rural China found that the central government's incentives are an important variable explaining why village cadres choose to implement some central policies instead of others, behaviour known as 'selective policy implementation' (O'Brien & Li, 1999). Cai (2004) points out that China's current political system provides both incentives and opportunities for the cadres to allocate or use resources in an irresponsible and wasteful manner due to the cadre promotion system in which Chinese local cadres must strive to impress higher level authorities through visible economic achievements.

What is the role of incentive structures in China's local environmental politics? This article analyses this question from three directions: the next section will explore the political incentives, which will be followed by an examination of the financial incentives and the morale incentives, respectively, in the fourth and fifth sections.

Methodologically, this article relies on interviews and document analysis. The primary data and materials consist of interviews, internal government documents, and files and local newspapers collected from fieldwork in Xiamen, Shihezi, and Ying Kou Economic and Technological Development Zone (YKETDZ) in 2007 and in Nantong in 2011. Secondary sources were collected from the Internet, newspapers, magazines, journals, and other public texts. Interviews were semi-structured in design. The interviewees were a variety of actors who influence environmental policy implementation at local levels, including government officials from different departments, everyday citizens, representatives from relevant polluting enterprises, journalists, members of grassroots non-governmental organizations (NGOs), environmental activists, scholars, and university students. One hundred and sixty-one people were interviewed either formally or informally: 59 in Shihezi; 54 in Xiamen; 35 in YKETDZ and 23 in Nantong. The author obtained access to conduct fieldwork in the four cities through the China Research Center for Comparative Politics and Economics (CCCPE) in Beijing. Most of the interviewees were reached through the local contacts of the CCCPE, so the interviews took different forms dependent on the specific setting and situation, including face-to-face in-depth interviews, focus group interviews, and informal interviews. The length of each interview varied from about 30 to 90 min, and was dependent on the interviewees' levels of interest in providing information. Because some of the informal interviews were more open and casual, the interviewees were more likely to speak frankly. However, they were often in a way that was too informal to be noted or quoted. In addition, while conducting fieldwork, I also witnessed observations similar to what Li and O'Brien found in their fieldwork that focused on popular protests in China: I found government officials, especially those in higher ranks who will retire or move to another position soon, they tended to be excellent subjects because of extensive experience and the fact that they were more likely to share substantial information as they are no longer attached to their current position (Li & O'Brien, 2008).[3]

My research deliberately focused on four cities of different administrative levels, which varied in terms of geography, social-economy situation, population, and environmental issues and reputation. Xiamen is a deputy-provincial level municipality in Fujian province, located on the southeastern coast of China, directly across the Taiwan Strait from Taiwan.[4] Its jurisdiction includes a land area of some 1565 km^2, with a sea area of over 300 km^2. At the end of 2011, the permanent population was about 3.53 million, and the GDP *per capita* was $10,947 US (Xiamen City Economic and Social Development Statistics Report, 2012). Xiamen is considered a 'model' city for environmental protection in China. It has won many environmental prizes and titles, including the 'National Model City for Environmental Protection', 'National Excellence in Tourism', 'National

Top 10 Most Livable Cities', 'National Habitat Award', and '2004 U.N. Habitat Award'.

However, local people are pessimistic about the changes in the condition of Xiamen's environment in recent years (Interview XM8, 2007). According to official documents internally published by Xiamen government, the water quality of Yuan Dang Lake and Jiu Long River has been seriously polluted; the marine environment has been damaged by household and industry sewage, fisheries, aquaculture, and land reclamation; and Xiamen is facing a serious groundwater shortage (Xiamen People's Government, 2007). Xiamen's air quality has worsened over the last couple of years due to increased industrial gas emissions, dust, and other pollutants from construction sites, and a tremendous increase in the number of vehicles (*Xiamen Daily*, 2009). In addition, research has also found that the transfer of pollution from downtown areas to suburbs and rural areas has become a serious environmental issue.

Xiamen made its mark in China's environmental history with the paraxylene ('PX') protest in 2007, which was the largest protest by number since Tiananmen.[5] A Taiwanese tycoon sought to invest 108 billion RMB to build a chemical factory in Xiamen's Haicang District, which would contribute to 80 billion to the local government's GDP (*Xiamen Daily*, 2007). Even though the construction of this factory was officially approved by the central and local governments, it was relocated after large-scale protests by local citizens. The peaceful protests composed of tens of thousands of residents and open public hearings concerning the project even attracted a large amount of international attention.[6]

Shihezi is a municipal-level city in the Xinjiang Uyghur Autonomous Region. It has a total area of 460 km^2, with a population of about 650,700 in 2010. In 2011, its GDP growth rate was 15.3% and its GDP *per capita* was $2400 US (Shihezi People's Government, 2012). Fifty years ago, it was a vast stretch of desert inhabited by only a handful of families. After 1949, a land-reclamation group—the Xinjiang Production and Construction Group (XPCG))—composed of demobilized soldiers and students, were sent there to build a new city in the Gobi Desert.[7] Unlike other cities in the Xinjiang Uygur Autonomous Region, Shihezi city is still under the leadership of the XPCG, and the majority of Shihezi residents are Han Chinese: ethnic minorities comprise only 5.4% of the population (Shihezi Statistics Bureau, 2007).

Before the turn of the millennium, it was a city dependent on agricultural and its light-industry output and was famous for its lush, healthy environment. Preventing desertification and sands storms via greening efforts in the desert is a major responsibility for the whole region. However, the city's development strategy has changed since then to emphasize heavy industry, especially the coal-chemical industry. The Megu Hu Reservoir, which receives almost all the city's industrial and household untreated wastewater, ranked among the top five most polluted reservoirs in the Xinjiang Uyghur Autonomous Region (*Xinjiang Daily*, 2008). Due to its location in the Gobi Desert, Shihezi has serious water shortage, and all household and industrial water is supplied from aquifers that are reduced 0.5 m each year (Shihezi Environmental Protection Bureau (EPB), 2011). Air quality has worsened in recent years due to the construction and expansion of four power plants and one coal-chemical plant, all of which lack proper desulphurization facilities and filters (*Legal Daily*, 2012).

YKETDZ is located in the southern part of Liaoning Province in northeast China. It is a district-level administration belonging to the Ying Kou Municipality, and it is the second-largest international seaport in northeast China.[8] As of 2011, its area was 268 km^2 with a population of 500,000; its GDP growth rate was 20%, and *per capita* GDP was about $4000 US (*YKETDZ Daily*, 2012). Over the past 20 years, YKETDZ has been transformed from a fishing village to an industrial city, which has inevitably led to significant environmental damage. The most serious case is the Bohai Sea, to which all sewage was discharged without treatment until 2010.[9] In early 2006, the Anshan Steel Company, one of the biggest steel companies in China, invested 22.6 billion RMB in a new seaside branch steel factory with a production capacity of 15 million tons per year. This investment not only composed more than half of the GDP reported by the local government, it also created serious air and water pollution problems (*Economic Observer*, 2007).

Nantong is a municipality level city in Jiangsu Province, East China. It is an important river port located on the northern bank of Yangtze River. In 2011, it had a total area of 8001 km^2 with a population of about 7,295,000 (Nantong People's Government, 2012). As of that year, its GDP growth rate was 12.1% and its GDP *per capita* was $8671 US (Nantong Statistics Bureau, 2012). Nantong's economy was traditionally based on agriculture and cotton textile production, but since 2000 it has become one of the fastest developing industry cities in the Yangtze River Delta (Interview NT3, 2011). The local residents worried about environmental degradation caused by the local government supported construction of several large polluting industries that were rejected by the governments of some neighbouring cities in the Yangtze River Delta due to environmental concerns (Interview NT15, 2011). Such worries recently boiled over during violent environmental protest by more than ten thousand residents, which forced the local government to cancel a planned industrial waste pipeline project for a Japanese-owned paper factory.

Political Incentives: Career Advancement and Bureaucratic Interest

Political incentives for policy implementation are those that provide the relevant actors and their institutions some form of political rewards in exchange for full implementation (McDermott, 2006). How does the central government politically motivate local cadres to implement environmental policies? Have the local cadres received effective political incentives from the central government to implement environmental policies? This section examines these questions through two perspectives: (1) Cadre Performance Evaluations and (2) bureaucratic interests of local governmental agencies. Ciqi Mei's research, which investigated the formal rules governing local chiefs' turnovers as well as actual past turnovers of the prefectural chiefs in Zhejiang and Hubei provinces during the reform era, found that the nature of political incentives embedded in the top-down cadre management system is 'incorrect' and unbalanced 'in that promotion criteria for local chiefs slant heavily to local chiefs' achievements (*zhengji*) in promoting local economic growth while their performance in other policy issues are neglected at large'—thereby encouraging local officials to produce distorted policy outcomes (Mei, 2009, p. 1). In their recent empirical research conducted in 12 Chinese counties and cities, Heberer and Senz suggested that while 'incentives alone are not sufficient for steering the behaviour of local cadres' to fulfill the national environmental policies, the cadre's political contract with the evaluation and target responsibilities has a positive effect on environmental policy implementation

(Heberer & Senz, 2011, p. 90). Based on the findings of my fieldwork, I argue that the central government provides political incentives that are often vague and internally compete and conflict with each other. Therefore, they failed to encourage not only local cadres, but also the local bureaucracies (agencies) whose task is to implement environmental policies; neither the local cadre's individual rewards of career advancement nor the implementation of institution's bureaucratic interests could be prioritized as the exchange of full environmental policy implementation.

One of the major goals of government officials is to seek and maximize their own political/administrative power. In a democratic system, obtaining votes is the major political incentive for those government officials who seek to be directly elected by local citizens. In the Chinese context, those responsible for implementing environmental policies are local cadres whose 'rational' choice is to seek promotion to higher positions in the political system as quickly as possible. Given the fact that China's local government officials are competitively selected and appointed by the Party Committee (*dangwei*), job promotion could be an effective political incentive to motivate local government officials to fully implement environmental policy (Heberer & Senz, 2011; Kostka, 2013). Within the current Party-State political system, the top-down Cadres Performance Evaluation System (CPES) is the major institutional mechanism for distributing political incentives to local cadres who faithfully implement Beijing's policies. The Organization Department of the Chinese Communist Party's (CCP) Central Committee (*zhong zubu*) determined the general regulations of the 'CPES', which specifies sets of targets (*kaohe zhibiao*) that local cadres must reach. Regulated by the central government's general principles, the upper level Organization Department of each Party Committee is responsible for evaluating the performance of cadres at the next-lower administrative level within a certain period in the given jurisdiction (Landry, 2008; Mei, 2009; Whiting, 2000).[10] The local cadres highly value positive evaluations since they are crucial for their career advancement (Interview SH3 & YKZ13, 2007).

While the specific contents of the 'CPES' set by the central government have been adjusted over time, this system consistently evaluates the local cadres based on five key aspects: 'morality' (*de*), 'capability' (*neng*), 'diligence' (*qin*), 'performance' (*ji*), and 'probity' (*lian*) (The Organization Department of the CCP, 2006, 2009). In 2007, six categories of quantitative targets were set to measure the cadres' performance in YKETDZ, namely: (1) economic development, mainly measured by the GDP growth rate and by revenue collection; 2) maintaining social stability, mainly measured by the number of 'mass incidents' (*qunti xing shijian*) and 'petition' (*shangfang*) cases; (3) development of education, science and technology, culture, health, and sports; (4) environment protection and population control; 5) public security; and (6) adherence to the Party's ideology.[11] Environmental protection remains loosely regulated in YKETDZ's cadre evaluation system. A failure to meet environmental targets, for instance, for the reduction of chemical oxygen demand (COD) and sulfur dioxide (SO_2) emissions would not get punished. Correspondingly, successful implementation of the government's soft environmental targets will not receive political rewards (YKETDZ People's Government, 2007). However, items that relate to economic growth and maintaining social stability, such as GDP growth rate, revenue growth, foreign and domestic investment, and the number of 'mass incidents' and 'petition' cases are treated as inflexible targets that must be met. Meeting these hard targets is an important basis for promotion and other political rewards. Failure to meet these hard targets negates any and all positive aspects of cadre performance (Interview SHZ55, YKZ1, & XM10, 2007).

The 'CPES' conveys information about which policies should receive priority and sends a signal of political incentives to local cadres. Researchers of Chinese rural politics find that local cadres are more likely to implement a clear and quantifiable 'hard target' with 'veto power' than a vague 'soft target' because the consequences of implementing a hard target are more decisive for local cadres' career advancement (Heimer, 2006; O'Brien & Li, 1999). Therefore, it is not surprising that, during my fieldwork, a few local government officials were more interested in environmental protection than economic development; on the contrary, officials made an effort to support industrial development at the expense of the environment and natural resources. When Mr Lifeng HE was promoted to the then Party Secretary of Xiamen, the central as well as Fujian province governments publicly stated in their performance evaluations that 'Comrade HE' is knowledgeable about the economy and very good at economic development', and his promotion was expected to 'contribute more to the economic development and investment growth in Xiamen' (*Xiamen Daily*, 2005). In an interview with a local Xiamen newspaper, he stated that since the primary investor in the 'PX' project was a Taiwan tycoon, the project was as one significant step towards implementing the 'Western Taiwan Straight Economic Zone' (*haixia xian jingji qu*) development plan; this plan is a Central Government strategy for improving cross-strait relations by 'buying' the hearts of Taiwan businessmen as well as contributing to local economic development (*Xiamen Daily*, 2006). According to his speeches and interviews, it was clear for him that implementation of the 'Western Taiwan Straight Economic Zone' development plan should have higher priority than implementation of environmental policies (*Xiamen Daily*, 2006, 2007). Although local people referred to him critically as 'crazy HE' due to his 'Great Leap Forward' style of developing the heavy industry at the cost of Xiamen's environment, and the 'PX' protesters called for him to step down, but he was still promoted to a higher position by the central government in 2009.

Kostka's empirical study on the career pattern at the provincial level EPB leaderships pointed out that the head of EPB's career background hardly provided them with incentives to properly implement environmental polices (Kostka, 2013). My fieldwork similarly found that EPB officials generally have less opportunity for promotion compared with their counterparts in the economic development-oriented bureaus, such as the Development and Reform Commission (DRC), Finance Bureau, Foreign Investment Bureau, and Construction Bureau. Not one of the top local government leaders had previous experience working at the EPB. More importantly, some of the EPB officials thought, in their words, that their careers are especially difficult because they have to worry about how implementing environmental policy might negatively impact their own career advancement. An EPB official at YKETDZ frankly admitted, 'We had to turn a blind eye to polluting industries since nobody would choose to implement central government policies at the risk of being fired'. When asked, 'Do you worry about the possibility of being fired if you fail to implement environmental policies?' he answered no, simply because this had never happened before (Interview YKZ27, 2007).

Environmental protection was not set as a hard target regulated by the 'CPES' in Xiamen, Shihezi, and YKETDZ at the time I conducted fieldwork in 2007. However, in the last 5 years, this situation has slightly changed because the central government has attempted to promote some environmental and energy targets into 'hard' targets with veto power that must be met by the local officials. China's 12th FYP, released in March of 2011, addressed growing environmental and energy concerns by setting seven binding targets, including energy intensity, carbon intensity (per unit of GDP), non-fossil fuels in the primary energy mix, major pollutants (COD and SO_2), forest coverage, water intensity (per unit of value-added industrial output), and farmland reserves. This new FYP is widely perceived as an ambitious blueprint for a Chinese green revolution. Some researchers have noted that energy and environmental targets in the 'CPES' have been combined at some localities, for instance, at Laixi in Shangdong province and Datong in Shanxi province (Heberer & Senz, 2011; Kostka & Hobbs, 2012). However, whether or not those abstract targets on paper can provide effective political incentives that will allow the targets to be translated into concrete environmental outcomes remains a big empirical question that requires careful examination at the local level.

My recent fieldwork in Nantong demonstrated that local government officials generally believe that the fulfilment of environmental targets is largely in conflict with the fulfillment of some other targets regulated by 'CPES', such as local GDP and revenue growth and the number of mass incidents (Interview NT 16 and NT17, 2011). While they are supposed to meet all of these hard targets, local governments usually prioritize some of the competing and conflict mandates. This is especially true because economic investment decisions affect some environmental targets more than other non-environmental targets.

It is widely assumed that the environmental policy implementation gap is largely caused by its nature of being soft targets, hence promoting environmental and energy goals to hard targets with veto power would be conducive to effective implementation. However, hard target setting and evaluation often results in the manipulation of statistical data. Statistical data manipulation has a long history in China, which reached its peak during the Great Leap Forward (Bernstein, 1984). This is also widely reported by Chinese national media, which call the phenomenon '"good" number produces officials, and "good" officials produce "good" number' (*shuzi chu guan, guan chu shuzi*).[12] Several researchers have pointed out that data manipulation has become routine for local cadres due to pressure from upper-level officials as well as their own interest in promotion and economic gain, so that is rooted in the institutional arrangement of China's administrative system (Cai, 2000; Chen, 2010). My fieldwork in Nantong supports their observations. When local cadres are faced with too many hard targets that they are unable to meet, they tend to manipulate the statistical data by over-reporting achievements, under-reporting failures, or using different statistical methods (Interview NT 19, 2011). As an official of the Nantong EPB put it: 'meeting targets is like playing a numbers game (*shuzi youxi*). It's impossible to meet all the binding targets, but still we have to meet those targets by whatever measures—which means we have to produce fake data'; and another explained that 'It won't be very hard to meet those targets once you know how to play the numbers game with central government' (Interview NT 20 and NT 19, 2011). The other reason why local cadres often report fake data is because the central government has no effective institutional mechanism for double-checking the statistical data. An official of the DRC in Nantong told me that the regulation and monitoring of energy intensity target by the National Development and Reform Commission (NDRC) is not very serious or strict; the measurement of energy intensity is very abstract, complicated, and has no national standards; the target can be met relatively easily because there are different methods for measuring energy consumption per unit of GDP, and local officials can use any method they prefer (Interview NT17, 2011). From a review of public official documents of the Central Government, it is clear that policy-makers in Beijing are familiar with the problem of data manipulation problem (State Council, 2012). Therefore, the problem is not simply that local cadres use fake data to cheat Central Government. Rather, the more important problem is the use of various methods of measurement, which has legitimized the local governments' behaviour since they can choose whichever method will best help them meet the requirements of hard environmental targets. In this sense, I argue that despite the promotion of environmental protection into a hard target regulated by the 'CPES', the political incentives are still not effective enough to encourage widespread of implementation of environmental policies at local levels. Rather, local cadres are politically incentivized to manipulate the environmental statistics.

The Central government has failed to provide effective political incentives not only for local cadres, but also for the local bureaucracies (agencies) whose task is to implement environmental policies. Policy implementation theory suggests that an implementation gap would be difficult to avoid if policy-makers assigned policy implementation tasks to too many agencies whose fundamental bureaucratic interests are to some degree in conflict with the goals of these policies (Matland, 1995). My research found that full implementation of environmental policies was not aligned with the bureaucratic interests of most agencies responsible for implementation, which was a result of the institutional defects inherently embedded in the system for the implementation of China's environmental policy at the local level. This is known as a 'fragmented authoritarianism' structure because the responsibility for policy implementation responsibilities are decentralized to various actors without the cooperation and coordination that would be necessary for successful implementation—instead, the decision-making power is highly centralized. Within this system, the responsibility can be visually represented by an inverted-pyramid structure, while the decision-making power can be represented by a pyramid structure (see Figure 1).

As the environmental administration agency, the local EPB is often assumed to have the highest responsibility and authority for environmental policy

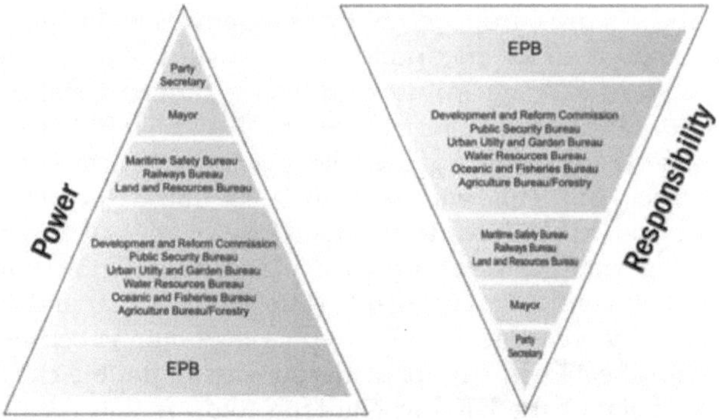

Figure 1. Pyramid-structured power, and the inverted pyramid of environmental responsibility. *Source:* The author.

implementation. However, the 'overall management and supervision' function of the EPB is restricted by its weak position (Jahiel, 1997; Lo & Tang, 2006). According to environmental laws, responsibility has been cut into pieces and decentralized to various actors. In total, there are more than 10 actors involved in the local environmental policy implementation system, namely, the Party Secretary, Mayor, EPB, DRC, Planning and Construction (Urban Utility and Garden) Bureau, Agriculture (Forestry) Bureau, Water Resources Bureau (WRB), Oceanic and Fishery Bureau, Maritime Safety Bureau, Land and Resources Bureau (LRB), Public Security Bureau (PSB), and the Railways Bureau.

The above listed agencies have been assigned not only environmental responsibilities, but also others that are often conflicting (see Table 1). Alford and Shen's (1998) research pointed out that environmental policy too often 'fails to anticipate the possibility that certain governments' interests, particularly those of depart-

Table 1. Environmental policies implementation agencies' environmental responsibility, non-environmental responsibility, and prioritizing interests

	Environmental responsibility	Non-environmental responsibility	Prioritizing interests
EPB	Pollution control, environmental monitoring, and ecological conservation		Environmental protection
DRC	Energy efficiency, climate change, low carbon city, master planning for ecological improvement, and environmental protection	Market regulation, overall social and economic planning, implementation of overall fiscal, financial, industrial, land, and social policies	Industry development and investment in public infrastructure development
Planning and construction (Urban Utility and Garden) Bureau	Sewage and waste treatment, city landscape, and building energy efficiency	Real estate industry regulation and public infrastructure planning and construction	Real estate industry development
Agriculture (Forestry) Bureau	Pesticide control, forestry protection, biodiversity, and ecological restoration	Agricultural productivity, regulation of rural enterprises, and rural poverty reduction	Agricultural output increase and rural enterprise development
WRB	Water resource protection and water quality control	Water infrastructure planning and construction, reservoir and dam construction and management, and flood and drought prevention	Water infrastructure construction, especially dams and reservoirs, collecting water resource fees
LRB	Land, mineral, and marine resource protection, natural disaster prevention	Land sale, collecting land and mineral resource fees, land requisition in rural areas, and demolition projects in urban areas	Collecting land and mineral resources fees

Source: The author.

ments with major economic responsibilities, might diverge sharply from those of local environmental protection officers' (p. 405). According to the law, the Water Resources Department oversees the protection of water resources as well as the development and usage of these resources. However, implementing water resource conservation and protection policies would challenge its bureaucratic interests, which are mainly built on water resource development like the construction of hydropower plants, reservoirs, drains, and dams (Mertha, 2008). The Development and Reform Committee (DRC) is in charge of energy and climate change issues, but its main interest is economic planning and regulation. As the most powerful organization in the State Council, the NDRC sets energy prices that are lower than the market value, because it must ensure that the market price for energy does not increase without administrative permission. This negatively influences its ability to assume its responsibility for developing and encouraging the use of renewable energy. In addition, their environmental responsibilities are often neglected in favour of concentrating on industry projects. Staff at the Shihezi DRC told me they have to help local investors to get cheaper prices for electricity and water to lower companies production costs; they cannot do much on environmental protection since the agency's primary goal is to attract investment from the industrial sector from which they gain better political performance (Interview SHZ9, 2007).

The Agriculture Bureau is responsible for agriculture pollution and for managing environmental problems in rural areas, but it is much more interested in promoting the development of agriculture industry and township enterprise. Agriculture Bureau officials in Shihezi City admitted that they had to only loosely implement fertilizer and pesticide environmental standards so that peasants could produce more agriculture products and increase their income—a priority for the Agriculture Bureau (Interview SHZ15, 2007). The Xiamen PSB only very recently began to enforce vehicle emissions regulations, which significantly contributed to air pollution in Xiamen. Not only did the PSB have no obligation to improve air quality, it benefited from vehicle growth due to the ability to collect more fees (Interview XM 44, 2007).

Financial Incentives: The Central Government's Environmental Budget

Successful policy implementation requires not only political incentives but also financial incentives for the implementers (Pressman & Wildavsky, 1984). Those responsible for implementing environmental policy at the local level do not simply highly value political incentives, but also calculate their own costs and benefits. The study on rural China found that 'village cadres' ability to serve village communities in China is influenced by both the financial resources that they have at their disposal and by the incentive structure under which they operate' (Kung *et al.*, 2009, p. 61). Do local governments and officials receive reasonable financial incentives from the Central Government for implementing environmental policies? This section tries to answer this question by examining the central government's public expenditures on environmental protection. My research suggests the central government does not financially motivate local governments to implement environmental policies due to its limited environmental budget. On the contrary, non-implementation or poor implementation did confer more financial rewards for local governments and officials.

Expenditure on environmental policy implementation at the local levels mainly pay for environmental management, pollution control, pollution reparation, natural resource protection, environmental infrastructure construction, green technology R&D, environmental education, and for EPBs' human resources and office expenses (Xiamen People's Government, 2007). Funding comes from several public and private channels, including: (1) the central government budget; (2) the local government budget; (3) bank loans; (4) industry investments; (5) NGOs and foreign aid (Interview YKZ26, 2007). No concrete data are available on how much of the public budget is intended for environmental purposes came from the central government before 2006, because 'environmental protection' expenditures have only been officially included in the central government's budget since that year. Until then, the central government had no fixed investments in environmental protection, and thus data on its financial contributions is full of uncertainty. Actually, only in 2006 did the central government announce that between 2005 and 2010 it would increase its overall environmental expenditures to 1.35% of GDP, including funds from the public (central and local governments) as well as the private sectors (SEPA, 2007). During the period of the 11th FYP (from 2006 to 2010), the total investment in environmental protection was around $700 billion, of which only 10% came from the central government's public budget (*People's Daily*, 2011).

Even though the central government's environmental expenditures provided funding for several major national-level and cross-regional environmental projects that are managed by the ministries of the central government, they are implemented by local governments. Examples of such projects include: the 'Energy Saving and Emissions Reduction Project', the 'Beijing and Tianjin Sandstorm Source Control Project', the 'Natural Forest Protection Project', the 'Returning Farmland to Forest (Grassland) Project', the 'Three Gorges Reservoir and Upper Stream Water Pollution Management Project', and the 'Three Rivers and Three Lakes Pollution Control Project'.[13] However, local governments would have to shoulder most of the financial burden if they tried to implement nationwide environmental reparation projects (Interview SHZ27, 2007).

Local governments may seek funding by applying in sequence to lower and then to higher administrative levels, but funding from the local level is a precondition for receiving funding from higher levels. The complex and time-consuming application procedures necessary for obtaining financial resources from higher levels of government or international aid via the central government dampen local governments' enthusiasm for acquiring such aid. Applying for funding requires a lot of time and energy because of the five administrative levels that must be negotiated and the general lack of cooperation among different government departments. Environmental projects are thus often delayed while waiting for funding from higher levels. For example, the best time to plant trees to prevent desertification in Shihezi is early spring, but the resources for this project often arrive from the central government during the summer or autumn. Implementation had to be delayed simply because the tree planting could not wait during the long and complicated bureaucratic procedure (Interview SHZ15, 2007). To get the environmental funding from the central government, building *guanxi* with the central government officials are crucial. Shihezi Tianye Company's recycling project got 0.4 billion RMB from the NDRC under the framework of 'recycling economy model enterprise encouragement'. However, a company insider admitted that successfully obtaining funding from the central government resulted mostly from top leaders in Beijing visiting their company a couple of times (Interview SHZ11, 2007).

In most cases, local governments have to finance a majority of their environmental expenditures, including EPBs' human resource and other standard office expenses, environmental management projects, environmental infrastructure construction, city greening projects, and so on. An official from the Forestry Bureau in Shihezi explained, 'Recently, we got more money from the central government than before, but it covered just 10 percent of the total expenditures for forestry in Shihezi' (Interview SHZ15, 2007). The head of the EPB at YKETDZ told me that they did not get any funding from the central government for pollution management and control in 2007, but did receive sufficient funding for human resources and other office expenses from the local government budget. He added, 'Our financial condition is better than that found in neighboring cities, where the EPBs' human resources and other office expenses cannot be covered by the local budget' (Interview YKZ6, 2007).

On the one hand, local governments face tremendous difficulties obtaining financial resources for environmental policy implementation; on the other hand, non-implementation or poor implementation would bring more material rewards for local governments and officials. Local government officials generally assumed that loosely implemented or non-implemented environmental policies could help them to attract more industrial investors in their jurisdiction. Following this logic, they tended to turn a blind eye to enterprises that create excessive pollution due to their tax contributions. The EPB is supposed to collect pollutant discharge fees from polluters and hand them over to the local Finance Bureau, which should spend the money on environmental reparation projects. However, official data showed that about 70.38 million RMB in pollution discharge fees went uncollected in Xinjiang in 2006, and only 8.2% of the fees were collected in Shihezi (Xinjiang EPB, 2007). A paper mill in Shihezi, which is a state-owned enterprise managed by the provincial-level government, has defaulted on its fees for about 5 years (Interview SHZ12, 2007). The Tianye chemical plant, also a provincial-level state-owned enterprise, refused to pay about 2 million RMB in pollution discharge fees from 2005 to 2007; even though it is a very rich enterprise, its leader knew that the company does not have to pay the pollutant discharge fees since it contributed 30% of the local GDP and therefore will not be punished just for environmental reasons (Interview SHZ56, 2007).

Moral Incentives: Local Officials Interpreting Environmental Policies

From a cognitive perspective, a key dimension of the implementation process is whether or not, and in what ways, implementing agents come to understand their practice, potentially changing their beliefs and attitudes in the process (Spillane *et al.*, 2002). Therefore, the moral incentive—teaching people to believe that a particular policy is appropriate and that loyal implementation is the right thing to do—is significant for achieving a desirable policy outcome (Honig, 2008). How did the local cadres frame, interpret, and construct the meaning of central environmental policy messages? Did implementers expect to improve their own sense of self-esteem and approval, or even to win admiration from others by successfully implementing environmental policies at the local level? On the other hand, did implementation failure result in a sense of guilt and condemnation, or even ostracism from the political system in China? My fieldwork suggests that it is difficult to find anything that can be identified as a moral incentive for environmental policy implementation in China. In general, local government officials shared two main perceptions shaping their interpretation of environmental policies handed down from the central government, both of which lead them to believe that the implementation of environmental policy would not bolster their self-esteem or generate admiration or even approval from others within the current political system; accordingly, they do not feel a sense of guilt upon failing to implement environmental policies.

First of all, local implementers generally believed that China's environmental policies are 'soft' laws with the character of high level conflict, ambiguity, and symbolism that are born to remain unimplemented. The aims and contents of policies promulgated by different policy-makers are in conflict to some degree, resulting from a lack of effective inter-organizational cooperation. For example, the EPB officials I interviewed in Shihezi and YKETDZ complained that Article 43 of the *Law of the People's Republic of China on Prevention and Control of Pollution from Environmental Noise*, enacted by the standing committee of the National People's Congress, stipulates that:

> Noise levels at the boundaries of newly constructed profit-making cultural and entertainment centers must be kept within the limits of environmental noise emission set by the state. No competent administrative department for culture may issue cultural operation licenses and no administrative department for industry and commerce may issue business licenses to centers that fail to keep emissions within these limits.

However, the regulations formulated by the State Administration for Industry and Commerce (SAIC)—the authority that actually issues and renews the business licenses—an environmental evaluation report is not among the governmental authorizations required by SAIC branches as a pre-requisite for issuing a business license (Interview YKZ27, 2007). This policy conflict between the two organizations means that many bars and clubs that produce serious noise pollution still hold legal business licenses, creating a significant environmental problem that local residents in Shihezi and YKETDZ complain about. It is these kinds of conflicts and lapses of enforcement that create the perception that the environmental laws are not meant to be enforced.

The goals of environmental policies and the means by which they are supposed to be implemented are, to a significant degree, ambiguous. Ambiguity of goals exists among China's stated economic and environmental goals and even in the nation's most important environmental law. The first article of the *Environmental Protection Law* states:

> This Law is formulated for the purpose of protecting and improving the People's environment and ecological environment, preventing and controlling pollution and other public hazards, safeguarding human health and facilitating the development of socialist modernization.

The vast majority of pollution is a direct result of industrial modernization: that has always been the case and is to be expected. Moreover, the ambiguous language in the text of environmental policies signals that Chinese environmental policies are intended to be ambiguous. Legal provisions are too often incomplete, inexplicit, or unclear. As other scholars have noted, 'environmental laws are framed in highly general and rhetorical terms. Significant elements of many major Chinese environmental measures seem more akin to policy statements than laws' (Wang & Liu, 1998, p. 440). The ambiguity creates a vacuum in which environmental laws lack real authority and the ability to impose penalties for failures to meet standards. Vague rhetoric such as 'shall' and 'encourage' are commonly used instead of more forceful and binding words such as 'must', which makes environmental laws difficult to enforce (Interview NT19, 2011).

Further hindering implementation and enforcement, the allocation of environmental responsibilities is also ambiguous. Several organizations are responsible for a single environmental issue without any clear division of labour, which in practice ultimately leads to a lack of accountability. This issue was supposed to be addressed by promotion of the State Environmental Protection Agency to the MEP in the new round of institutional reforms initiated by the State Council in March 2008, however, the environmental responsibilities are still fragmentally decentralized to various ministries. A third area of ambiguity concerns the applicability of environmental policies—it is unclear which environmental policies should be applied in which circumstances. As written, the environmental policies are too abstract, with too few measurable standards or objectives to actually be put into practice. For instance, the Law of the PRC on the *Prevention and Control of Atmospheric Pollution* states that:

> Newly built or expanded thermal power plants and other large or medium-sized enterprises that discharge sulfur dioxide at amounts exceeding the prescribed standards must install supporting facilities for desulphurization and dust removal or take other measures to control the discharge of sulfur dioxide or dust.

However, the problem confusing policy implementers is that no regulations precisely define what constitutes a medium-sized or large enterprise (Interview SHZ8, 2007).

Policy analysis theory suggest that symbolic policy plays an important role in confirming and emphasizing important values or principles in a context in which only policy-makers are aware of any gaps that exist between these values and their practice (Gustafsson, 1983). Theoretically, it seems implausible that high levels of ambiguity and conflict would exist in China's environmental policies if policy-makers in Beijing genuinely have a strong and serious intention to implement those policies at local levels. However, government decision-makers may know that discrepancies exist between the publicly stated political goal of environmental protection and the more complex and difficult realities of implementing measures to achieve this goal in China, and may, therefore, be making symbolic policies. As one official stated in an interview:

> Central government leaders are not stupid, they are very smart. They may assume from the beginning that full implementation of their environmental policies would negatively influence GDP growth and employment, from which they derive their legitimacy and job security. (Interview XM21, 2007)

In my fieldwork, I found that local officials tended to believe that, under the circumstances discussed above, the Central Government was making symbolic environmental policy instead of 'hard' policy. By doing this, China's central government has successfully presented its major concerns regarding environmental issues to the world and to its citizens. While several innovative ideas and policies that claim to deal with the environmental crisis are rhetorically brilliant, they are hardly put into practice. Such examples include the 'Green GDP', 'China's 21st Century Agenda' and the 'National Resolution on Global Climate Change' (Stenhardt & Jiang, 2007). The central government has already received a tremendous outpouring of sympathy and understanding from domestic and international society due to the comprehensive and modern environmental policy framework it has presented. Local governments, on the contrary, are blamed for not effectively implementing this policy.

Although symbolic policy may be helpful in strengthening policy-makers' legitimacy in Beijing, it creates conflicts for policy implementation at local levels. In practice, considerable overlap exists between real and symbolic decisions, and the complexity of the policy-making process introduces a danger that symbolic policy ingredients will be intermingled with real meta-policy. Local officials have to devote significant time and other resources to interpreting and understanding central-level policies. They value the cadre performance evaluation system not only because their promotion or demotion is largely determined by it, but also because the evaluation allows them to differentiate the central government's symbolic policies from their real policies, allowing the local officials to implement them selectively. Once a certain policy has been interpreted as being symbolic, local officials will not put forth much effort to implement it. In addition, they will not feel guilty for this lack of implementation (Interview SHZ53 2007; Interview NT14, 2011).

A second perception influencing how local implementers interpreted environmental policy is a sense that leaders at the central levels do not care about environmental protection very much, since the current guidelines signal that environmental policies should serve economic development above all else. The cognitive consensus between the central and local government officials is that economic development deserves to be the top priority. A survey found that even EPB officials were very much in favour of the government emphasizing economic growth and believed that implementation of aggressive environmental protection policies would hinder the nation's economic growth (Tong, 2007). According to the survey by the Chinese Academy of Social Science (CASS) in 2007, Chinese government officials do generally believe that environmental protection is important, and that effectively implementing environmental policies at the local level is the right thing to do. However, they also consider other policy areas, such as job creation and economic development, as a priority (CASS, 2008). The findings from my fieldwork also support the above survey results. Most of the EPB officials interviewed favoured economic development, saying things such as, 'Without money and technology it would be impossible to explore any pollution clean-up projects' (Interview YKZ28, 2007); and 'We have to sacrifice the environment a little bit for economic development' (Interview NT18, 2007). However, differences were found among the cities studied here. Some local government officials thought that Xiamen had already crossed over into an economically developed phase and should not continue to develop at the cost of the environment (Interview XM36, 2007). Some stated that it is time to pay more attention to pollution cleanup, as Western countries have done (Interview XM30, 2007). However, almost all the interviewees in Shihezi and YKZ still believed that environmental costs are necessary in order to develop industry.

This belief that the environment has to be sacrificed for China's economic progress, along with a perception that the environmental policy is in conflict, ambiguous, and symbolic provides those responsible for policy implementation with no moral incentive, and hence, no feeling of guilt for their implementation failures.

Conclusions

One of the obvious paradoxes in China's environmental politics is the large gap between the central government's policies and their implementation outcomes at local levels. Most of the literature agrees that addressing this implementation gap is crucial for better and more effective environmental governance in China (Carter & Mol, 2007). However, the existing literature provides two different ways of understanding the role of the local government in China's environmental politics. One school blames local governments for being the biggest obstacle to environmental policy implementation for two reasons: (a) they are development-oriented at a cost to the environment and (b) the central government has decentralized too much power to control local governments, while maintaining very weak control (Economy, 2004; Lo & Tang, 2006). The other school argues that in terms of policy innovation and environmental activism local 'environmental states' are emerging; local governments have initiated many good practices to meet the central government's comprehensive set of mandates as well as to build their own green reputations (Li *et al.*, 2011).[14] This article aims to bring the central government back to the centre of analysis by examining the effects of incentive structure on the policy implementation gap in China's local environmental politics.

Drawing on fieldwork and documentary analysis, my research demonstrates that the incentives set by the central government for environmental policy implementation at the local level are perversely structured, meaning that the central government provides more incentives for local government's non-implementation or poor implementation of its environmental policies than for full implementation. The central government's failure to encourage—politically, financially, as well as morally—local government officials to appropriately implement environmental policies can partly explain the implementation gap in China's local environmental politics. Unless central government takes significant measures to address the system of perverse incentives embedded in the overall structure of China's environmental politics, the environmental policy implementation gap cannot be overcome by the efforts at the local level. From this perspective, the central government, rather than local governments, should be more responsible for addressing the policy implementation gap and cleaning up China's environment.

Blaming local governments that fail to implement environmental policy, while praising the central government for being a morally innocent policy-maker as it passes attractive-looking policies, is very common but is also illogical. There is no essential conflict of interest between the central and local governments in the area of environmental policy. Both want GDP growth more than anything else because it benefits them directly and because, they believe, it leads to social stability. Local government officials are hypnotized by economic growth to such a degree that they turn a blind eye to pollution, a situation that is mostly due to the current perverse incentive structure set by the central government. This structure makes them believe that the central and local governments are in cognitive agreement, and that environmental problems are not a priority in the policy agenda when compared with GDP growth and maintaining social stability. Due to this lack of emphasis, full implementation of the environmental policies would not bring them political, financial, or moral benefits, which thereby ultimately leads to the implementation gap.

Acknowledgements

This research is supported by the Haniel Foundation, US-China Institute of University of Southern California and the Fundamental Research Funds for the Central Universities, and the Research Funds of Renmin University of China (grant number:11XNK009). An earlier version of this paper was presented at the 2012 Annual Conference of Association for Asian Studies (AAS) in Toronto, March 2012. I am very grateful for the thoughtful comments from Arthur Mol, Genia Kostka, Amir Hampel, Kyoung Shin, Nicholas Holtkamp and the two peer reviewers.

Notes

1. For a comprehensive literature review on Chinese environmental politics, see Wu (2009).
2. Here the policy implementation gap refers to the differences between the central level's environmental policies written in text and actual implementation outcomes produced at local levels. The 'policy implementation gap' is more or less fluid and depends on the interaction between the goals of policy-making and their implementation outcomes in practice: the bigger the differences between policy goals and actual outcomes, the bigger the implementation gap.
3. Interviewee SHZ31 and Interviewee XM21 are two typical examples.
4. As one of the first of four special economic zones in China, Xiamen is one of the deputy-provincial levels municipalities that have independent status in state economic planning, and also has provincial-levels autonomy and local legislative power in economic administration.
5. PX is considered highly toxic and has been linked to birth defects in animals. Short-term exposure can cause eye, nose, or throat irritation in humans, according to the US Centers for Disease Control and Prevention. Chronic exposure can affect the central nervous system and may cause death. Benzene is a by-product of PX and is a toxic chemical that occurs naturally in the environment and is also used in a wide range of products. Exposure can result in a range of side effects, which can be acute or chronic, and even deadly. Long-term or high-levels exposure to benzene can result in a number of ailments. This chemical has been labelled a class 'A' carcinogenic by the US Environmental Protection Agency.
6. For the representative international media's reports, see 'China city gets the (text) message', in *Los Angels Times*, print edition A-3. 1 June 2007; 'Mobilised by mobile', In *Economist* print edition, 21 June 2007; China bows to public over chemical plant: environmental protesters score landmark victory, *Nature*, 451(117), 2008, published online 9 January 2008.
7. XPCG is a special organization with the functions of administration, economy, and military in its jurisdiction.
8. Since 1992, it has been a national-levels economic and technology development zone with the administrative power of a vice-municipality.
9. No sewage and waste treatment plants were available until 2010 in YKETDZ.
10. More discussion on China's cadre evaluation system, see Whiting (2000) and Edin (2003).
11. '*Guanyu Yingkou Shi Ganbu Kaohe Zhibiao Tixi de Guiding*' [The Regulation of Yingkou City Cadre Performance Evaluation Indicators System], documents collected in the fieldwork, 2007.
12. For more media coverage, see the on-line discussion on the website of *People's Daily*. Available at http://opinion.people.com.cn/GB/40604/3648587.html (accessed 20 July 2012).
13. For more details, see '2008 *Zhongyang Caizheng Yusuan Baogao Caoan*' [Report of 2008 Central Budget Draft Plan]. Available at http://www.npc.gov.cn/huiyi/dbdh/11/2008-03/06/content_1406960.htm (accessed 6 March 2010).

14. For more media coverage on this topic, see 'How to Spot a Fake Eco City?'. Available at http://www.chinadialogue.cn/article/show/single/ch/4488-How-to-spot-a-fake-eco-city (accessed 27 September 2011).

References

Alford, W. P. & Shen, Y. (1998) Limits of the law in addressing China's environmental dilemma, in: M.B. McElroy, C. P. Nielsen & P. Lydon (Eds), *Energizing China, Reconciling Environmental Protection and Economic Growth*, pp. 405–429. (Cambridge, MA: Harvard University Press).

Bernstein, T. P. (1984) Stalinism, famine, and Chinese peasants – Grain procurement during the Great Leap Forward, *Theory and Society*, 13(3), pp. 339–377.

Cai, Y. S. (2000) Between state and peasant: Local cadres and statistical reporting in rural China, *China Quarterly*, (163), pp. 783–805.

Cai, Y. S. (2004) Irresponsible state: Local cadres and image-building in China, *Journal of Communist Studies and Transition Politics*, 20(4), pp. 20–41.

Carter, N. T. & Mol, A. J. (Eds) (2007) *Environmental Governance in China* (London: Routledge).

CASS (2008) *2007 Nian Zhongguo Gongzhong Huanjing Yishi Diaocha Baogao* [2007 China General Public Environmental Survey]. Available at http://www.chinaceap.org/download/8.pdf (accessed 9 June 2010).

Chen, X. (2010) State-generated data and the study of contentious politics in China, in: A. Carlson, M. Gallagher, K. Lieberthal, M. Manion (Eds) *Chinese Politics: New Methods, Sources and Field Strategies*, pp. 15–32 (New York: Cambridge University Press).

Clark, P. & Wilson, J. (1961) Incentive systems: A theory of organizations, *Administrative Science Quarterly*, 6(3), pp. 129–166.

Economic Observer (2007) Gangyu Gang de Jiehe: Yingkou Kaifa Qu 15 Nianhou Zhaodao le Fazhan Fangxiang [Steel and Port Mixture: YKETDZ Found Its Development Direction 15 years later], 6th ed., May 20.

Economy, E. C. (2004) *The River Runs Black: The Environmental Challenge To China's Future* (Ithaca, NY: Cornell University Press).

Economy, E. C. (2007) The Great Leap Backward, *Foreign Affairs*, September/October, 86(5), pp.38–59.

Edin, M. (2003) State capacity and local agent control in China: CCP cadre management from a township perspective, *China Quarterly*, (173), pp. 35–52.

Greenpeace (2011) *Greenpeace Insights & Commentary: China's 12th Five-Year Plan*. Available at http://www.greenpeace.org/eastasia/Global/eastasia/publications/reports/others/2011/greenpeace-commentary-twelfth-five-year-plan.pdf (accessed 7 July 2012).

Gustafsson, G. (1983) Symbolic and pseudo policies as responses to diffusion of power, *Policy Sciences*, 15(3), pp. 269–287.

Hahn, R. W. & Stavins, R. N. (1992) Economic incentives for environmental protection: Integrating theory and practice, *The American Economic Review*, 82(2), pp. 464–468.

Heberer, T. & Senz, A. (2011) Streamlining local behavior through communication, incentives and control: A case study of local environmental policies in China, *Journal of Current Chinese Affairs*, 40(3), pp. 77–112.

Heimer, M. (2006) The cadre responsibility system and the changing needs of the party, in: K. E. Brodsgaard and Y. N. Zheng (Eds) *The Chinese Communist Party in Reform*, pp. 122–138 (London: Routledge).

Honig, M. I. (2008) District central offices as learning organizations: How sociocultural and organizational learning theories elaborate district central office administrators' participation in teaching and learning improvement efforts, *American Journal of Education*, 114, pp. 627–664.

Jahiel, A. R. (1997) The contradictory impact of reform on environmental protection in China, *China Quarterly*, (149), pp. 81–103.

Kostka, G. (2013) Environmental protection bureau leadership at the provincial level in China: Examining diverging career backgrounds and appointment patterns, *Journal of Environmental Policy and Planning*, forthcoming.

Kostka, G. & Hobbs, W. (2012) Local energy efficiency policy implementation in China: Bridging the gap between national priorities and local interests, *China Quarterly*, (211), pp. 765–785.

Kung, J., Cai, Y. S. & Sun, X. L. (2009) Rural cadres and governance in China: Incentive, institution and accountability, *The China Journal*, (62), pp. 61–77.

Landry, P. F. (2008) *Decentralized Authoritarianism: The Communist Party's Control of Local Elites in the Post-Mao Era* (New York: Cambridge University Press).

Legal Daily (2012) Xinjiang Huagong Qiye Wuran Muqinhe Xiangguan Bumen Jianer Buguan [The Mother River is Polluted by Chemical Plant in Xinjiang, Government's Supervision is Ineffective]. Available at http://big5.xinhuanet.com/gate/big5/www.xj.xinhuanet.com/2012-08/30/c_112902311.htm (accessed 9 December 2012).

Li, L. J. & O'Brien, K. (2008) Protest leadership in rural China, *The China Quarterly*, (193), pp. 1–23.

Li, Y., Bo, M. & Lang, G. (2011) The local environmental state in China: A study of county-level cities in Suzhou, *The China Quarterly*, (205), pp. 115–132.

Lieberthal, K. (1997) *China's Governing System and Its Impact on Environmental Policy Implementation*, China Environment Series (Washington, DC: Woodrow Wilson).

Lo, C. W. H. & Tang, S. Y. (2006) Institutional reform, economic changes, and local environmental management in China: The case of Guangdong province, *Environmental Politics*, 15(2), pp. 190–210.

Lopes, C. & Theisohn, T. (2004) *Ownership, Leadership and Transformation: Can We Do Better for Capacity Development?* (New York: Earthscan Publications).

Mathauer, I. & Imhoff, I. (2006) Health worker motivation in Africa: The role of non-financial incentives and human resource management tools, *Human Resources for Health*, 4, p. 24.

Matland, R. E. (1995) Synthesising the implementation literature: The ambiguity-conflict model of policy implementation, *Journal of Public Administration*, 5(2), pp. 145–174.

McDermott, K. A. (2006) Incentives, capacity, and implementation: Evidence from Massachusetts education reform, *Journal of Public Administration Research and Theory*, 16(1), pp. 45–65.

Mei, C. Q. (2009) Brings the politics back in: Political incentive and policy distortion in China, PhD dissertation, Maryland University, USA.

Mertha, A. C. (2008) *China's Water Warriors: Citizen Action and Policy Change* (Ithaca, NY: Cornell University Press).

Mol, A. P. J. (2011) China's ascent and Africa's environment, *Global Environmental Change: Human and Policy Dimensions*, 21(3), pp. 785–794.

Nantong People's Government (2012) Nantong Gaikuang [Brief Introduction of Nantong], website of Nantong Government. Available at http://www.nantong.gov.cn/ (accessed 9 December 2012).

Nantong Statistics Bureau (2012) *Nongtong Tongji Nianjian 2011* [Nantong Statistics Year Book 2011] (Beijing: China Statistics Press).

O'Brien, K. & Li, L. (1999) Selective policy implementation in rural China, *Comparative Politics*, 31(2), pp. 167–186.

OECD (2007) *OECD Environmental Performance Reviews China*, OECD Publishing. Available at http://www.oecd.org/document/47/0,3343,en_33873108_36016481_37809647_1_1_1_1,00.html (accessed 2 May 2011).

Pan, Y. (2006) Using green GDP to support scientific development view: An interview with Pan Yue', website of MEP. Available at http://panyue.mep.gov.cn/zyhd/200907/t20090708_154443.htm (accessed 25 July 2012).

Pennington, M. (1999) Free market environmentalism and the limits of land use planning, *Journal of Environmental Policy & Planning*, 1(1), pp. 43–59.

People's Daily (2011) Shiyiwu Qijian Huanbao Touru Dafu Tigao [The Environmental Protection Investment Jumped during the 11th FYP Period], January 14, 2011, edition 16.

Pressman, J. & Wildavsky, A. (1984) *Implementation: How Great Expectations in Washington are Dashed in Oakland* (Berkeley, CA: University of California Press).

SEPA (2007) Huanbao Caizheng Yusuan de Dazhan [The Development of Environmental Protection Budget], official document collected in fieldwork.

Shihezi People's Government (2012) Shihezi Jianjie [Brief Introduction of Shihezi City], website of Shihezi government. Available at www.xjshz.gov.cn (accessed 15 June 2012).

Shihezi Statistics Bureau (2007) *Shihezi Tongji Nianjian 2007* [Shihezi Statistics Year Book 2007] (Beijing: China Statistics Press).

Shihezi Environmental Protection Bureau (2011) Shihezi Huanjing Zhiliang Gongbao 2010 [Shihezi Environmental Quality Report]. Available at http://www.xjepb.gov.cn/tabid/787/InfoID/348153/frtid/41/Default.aspx (accessed 22 June 2011).

Spillane, J. P., Reiser, B. J. & Reimer, T. (2002) Policy implementation and cognition: Reframing and refocusing implementation research, *Review of Educational Research*, 72(3), pp. 387–431.

State Council (2012) Guanyu Zuohao Jieneng Jianpai Tongji Gongzuo de Tongshi [The Notice on Improving the Statistics of Energy Saving and Emission Reduction], official document collected in fieldwork.

Stenhardt, H. C. & Jiang, Y. H. (2007) The politics of China's green GDP, *Journal of Current Chinese Affairs*, 36(5), pp. 25–39.

Tang, S. Y., Tang, C. P. & Lo, C. W. H. (2005) Public participation and environmental impact assessment in mainland China and Taiwan: Political foundations of environmental management, *The Journal of Development Studies*, 41(1), pp. 1–32.

The Organization Department of the CCP (2006) Tixian Kexue Fazhanguan Yaoqiu de Difang Dangzheng Lingdao Banzi he Lingdao Ganbu Zonghe Kaohe Pingjia Shixing Banfa [The Experimental Policy on Local Political Leaders and Cadres Comprehensive Evaluation System Which Embody the Requirement of Scientific Development Outlook], internal document collected in the fieldwork.

The Organization Department of the CCP (2009) Difang Dangzheng Lingdao Banzi he Lingdao Ganbu Zonghe Kaohe Pingjia Banfa [The Policy on Local Political Leaders and Cadres Comprehensive Evaluation System], internal document collected in the fieldwork.

Tilt, B. (2007) The political ecology of pollution enforcement in China: A case from Sichuan's rural industrial sector, *The China Quarterly*, (192), pp. 915–932.

Tong, Y. Q. (2007) Bureaucracy meets the environment: Elite perceptions in six Chinese cities, *China Quarterly*, (189), pp. 100–121.

Ulleberg, I. (2009) *Incentive Structures as a Capacity Development Strategy in Public Service Delivery*, Published by International Institute for Educational Planning, UNESCO. Available at http://www.iiep.unesco.org/fileadmin/user_upload/Cap_Dev_Rethinking/pdf/CapDev_Incentives.pdf (accessed 1 July 2012).

Van Rooij, B. (2006) Implementation of Chinese environmental law: Regular enforcement and political campaigns, *Development and Change*, 37(1), pp. 57–74.

Van Rooij, B. & Lo Wing-Hung, C. (2010) Fragile convergence: Understanding variation in the enforcement of China's industrial pollution law, *Law & Policy*, 32(1), pp. 14–37.

Wang, H. C. & Liu, B. J. (1998) Policymaking for environmental protection in China, in: M. B. McElroy, C. P. Nielsen & P. Lydon (Eds) *Energizing China, Reconciling Environmental Protection and Economic Growth*, pp. 371–404 (Cambridge, MA: Harvard University Press).

Watts, J. (2010) *When A Billion Chinese Jump: How China Will Save Mankind – Or Destroy It* (New York: Scribner).

Whiting, S. H. (2000) *Power and Wealth in Rural China: The Political Economy of Institutional Change* (Cambridge: Cambridge University Press).

World Bank & SEPA (2007) *Cost of Pollution in China: Economic Estimates of Physical Damages*. This publication is available online at http://siteresources.worldbank.org/INTEAPREGTOPENVIRONMENT/Resources/China_Cost_of_Pollution.pdf (accessed 7 December 2012).

Wu, F. S. (2009) Environmental politics in China: An issue area in review, *Journal of Chinese Political Science*, 14, pp. 383–406.

Xiamen Daily (2005) Renming Helifeng Tongzhi wei Xiamen Shiwei Shuji de Jueding [Decision Concerning Appointment of Comrade HE Lifeng as the Party Sectary of Xiamen], 1st ed., May 21.

Xiamen Daily (2006) Luoshi Kexue Fazhanguan Jianshe Haixi Jingjiqu [Implementing Scentific Development Outlook, Construction the Western Taiwan Straight Economic Zone], 2nd ed., October 22, 2012.

Xiamen Daily (2007) Xiamen Caizheng Buru Kuaisu Zengzhang qi [The Government Revenue Increasing Rapidly in this Period], 1st ed., January 1.

Xiamen Daily (2009) Guanyu Gaishan Xiamen Huanjing Kongqi Zhiliang de Duice he Jiangyi [Suggestions on How to Improve Xiamen's Environmental and Air Quality], 3rd ed., August 5.

Xiamen People's Government (2012) 2011 Nian Xiamen Shi Guomin Jingji he Shehui Fazhan Tongji Gongbao [2011 Xiamen City Economic and Social Development Statistics Report]. Available at http://cn.chinagate.cn/infocus/2012-07/30/content_26061268.htm (accessed 7 December 2012).

Xiamen People's Government (2007) Shui Ziyuan Kechixu Fazhan Cujin Xiamen de Kechixu Fazhan [Water Resources Sustainability Promotes Xiamen's Sustainable Development], Internal official documents collected in the fieldwork.

Xiamen People's Government (2007) Xiamen Huanbao Touru Qingkuang de Shuoming [The Introduction of Environmental Protection Investment in Xiamen], Internal official documents collected in the fieldwork.

Xinjiang Daily (2008) Xinjiang Huanbao Mianlin de Nanti [The Difficulties of Environmental Protection in Xinjiang], 2nd ed., November 8.

YKETDZ Daily (2012) 2011 Nian Woqu Shixian Jingji Kuayueshi Fazhan [The Leap-Forward Style Economic Development in 2011], 1st ed., January 16.

YKETDZ People's Government (2007) Yingkou Kaifaqu Lingdao Ganbu Zonghe Kaohe Pingjia Zhibiao Tixi [YKETDZ Cadres Comprehensive Evaluation Indicators System], Internal document collected in the fieldwork.

Zhou, S. X. (2008) Zhou Shengxian zai 2008 Quanguo Huanjing Zhifa Gongzuo Huiyi Shangde Jianghua [Zhou Shengxian's Speech in the Environmental Law Enforcement Meeting 2008], website of Ministry of Environmental Protection. Available at http://www.zhb.gov.cn/info/ldjh/200804/t20080425_121679.htm (accessed 8 July 2012).

Interviews

Interview NT3, 2011; Interview NT14, 2011; Interview NT15, 2011; Interview NT17, 2011;

Interview NT19, 2011; Interview NT20, 2011;

Interview SHZ3, 2007; Interview SHZ8, 2007; Interview SHZ9, 2007; Interview SHZ11, 2007;

Interview SHZ12, 2007; Interview SHZ15, 2007; Interview SHZ27, 2007; Interview SHZ53, 2007; Interview SHZ55, 2007; Interview SHZ56, 2007;

Interview XM8, 2007; Interview XM10, 2007; Interview XM21, 2007; Interview XM30, 2007;

Interview XM36, 2007; Interview XM44, 2007;

Interview YKZ1, 2007; Interview YKZ6, 2007; Interview YKZ13, 2007; Interview YKZ26, 2007; Interview YKZ27, 2007; Interview YKZ28, 2007.

Environmental Protection Bureau Leadership at the Provincial Level in China: Examining Diverging Career Backgrounds and Appointment Patterns

GENIA KOSTKA

Department of Economics, Frankfurt School of Finance and Management, Frankfurt, Germany

ABSTRACT *This paper analyses the career backgrounds of local government officials in provincial Environmental Protection Bureaus (EPBs) in China and explains the appointment patterns of Chinese EPB bureaucrats. Using biographical information of provincial EPB heads and drawing on fieldwork conducted in Shanxi Province and Inner Mongolia, this paper finds that only one-fourth of the provincial EPB heads were promoted through the bureau ranks within the EPBs, while the remaining three-fourths were appointed from positions outside the environment field. Further, nearly all EPB heads' professional backgrounds and associated networks can be clearly categorized as environmental, business, provincial government, or local government oriented. This paper delineates these four types of Chinese EPB leaders and explains why an awareness of the different professional orientations is critical to understanding environmental regulation and protection in China. These findings have implications for inferring the unique characteristics of a province's EPB leadership, the implementation capacities of provincial EPBs, and the appointment preferences of provincial leaders.*

Introduction

Previous research on China has focused on the institutional limitations of local Environmental Protection Bureaus (EPBs). These include studies in Guangdong (Lo & Tang, 2006; Sinkule & Ortolano, 1995), Shandong (Ma & Ortolano, 2000), Sichuan (Van Rooij, 2003), and Liaoning (Bauer, 2006). It seems uniformly acknowledged that EPBs are often understaffed, underfunded, and lacking in the necessary formal power to implement many environmental tasks assigned to them. What is implicit in these studies is that if these EPBs were granted the institutional capacity to function efficiently and authoritatively in their localities, then they would be much more likely to achieve their environmental protection objectives.

This paper evaluates the implementation capacity of EPBs through an alternative approach. It evaluates institutional capacity by analysing the backgrounds of current provincial EPB heads and explains why some EPB heads are selected over others.[1] The EPB heads are often authoritative figures within the bureaus and have an overriding influence over the internal workings of bureaus' environmental protection efforts (Van Rooij, 2003). However, the EPB heads are themselves only one small part of an intricate network of government agencies, which rely on inter-departmental coordination and consensus-building to achieve their bureau objectives (Sinkule & Ortolano, 1995). Bureau heads are promoted into their positions by provincial leaders and obey the explicit orders and implicit priorities of these leaders. As EPB heads are authoritative and influential figures within an EPB (Bauer, 2006; Van Rooij, 2003), the institutional capacities, relative to each other, of provincial EPBs can be to a large extent determined by the priorities and objectives of the current EPB heads. The overall capacities, or willingness, of the EPB heads to undertake strict environmental protection measures are heavily influenced by their relationship with their superiors in the provincial government and their colleagues in other bureaus.

While previous research has analysed leaders' backgrounds, it has mainly studied national and provincial level leaders, who include provincial Party secretaries and mayors (Huang, 2002; Li, 2010; Mei, 2009). Very little of this literature depicts provincial bureau heads as active agents. This paper aims to fill this gap, providing insight into career trajectories and biographical data and the possible implications of these factors on environmental protection performance. Who are these departmental heads and what light can their backgrounds shed on the appointment preferences of provincial leaders and the workings and capacities of provincial EPBs? Do EPB heads share similar demographic characteristics and professional experiences across China? Specifically, an analysis of the EPB heads' biographical information helps to achieve four important objectives: (1) reveal the career patterns of provincial leading bureaucrats in charge of environmental policy implementation; (2) exhibit the professional networks of these bureaucrats; (3) assess how career backgrounds and professional networks influence the implementation capacity and independence of local EPBs; and (4) infer the appointment preferences of provincial Party committees and organizational departments in charge of appointing provincial EPB heads.

The analysis of EPB departmental heads' career patterns and institutional backgrounds demonstrates the variety of non-environmental career backgrounds, concurrent affiliations, and professional networks. This paper argues that EPB leaders are highly embedded in their locality and in the local Party and government organizations. This embeddedness is a double-edged sword. While it improves the implementation capacity of EPB leaders, it can also dilute environmental protection incentives. This implies that there is a delicate balance between professional networks that improve implementation capacity and relations that paralyse efforts to improve environmental protection in China. The study further shows that EPB leaders may be easily categorized by their career backgrounds. The distinct characteristics of these 'types' suggest that provincial leaders make important appointment decisions among clearly differentiated candidates. This may have implications for inferring the appointment preferences of provincial leaders. The most desirable candidate for the EPB head position will likely be an official whose assumed networks and incentives will be favourably aligned with provincial priorities and with parochial interests of the provincial leaders. Appointing provincial leaders are more likely to opt for an EPB head with the strength to help them to meet their individual mandatory or 'hard' (*ying xing* 硬性) targets under the cadre responsibility system, among which provincial leaders identified some targets that are more difficult to meet than others. Thus, provincial leaders selecting the most able EPB head may not behave altruistically, but, ultimately, in their own self-interest.

The analysis draws from biographical information of provincial EPB heads and interviews undertaken during 2010 and 2011. The author collected biographical data for departmental heads of provincial EPBs holding office as of February 2011 in China's 31 provinces (including centrally administered municipalities and autonomous regions). The majority of biographies of EPB heads are published on government websites. Where information was not available, inconsistent, or incomplete, additional information was sought in newspaper articles, government reports, or the *baidu* online encyclopaedia. The analysis also draws on 71 semi-structured interviews conducted in Shanxi and Inner Mongolia in 2010 and 2011. While Shanxi, a province under external pressure to improve air pollution, selected a candidate with the skills and credentials to effectively implement environmental mandates, the still-developing Inner Mongolia appointed a candidate who can balance economic growth and environmental protection concern. Interviews with local organization departments clarified how EPB heads get appointed and demonstrate the important role played by Party committees and organizational departments. Interviews with EPB heads and lower level EPB officials provided an understanding of the responsibilities assigned to EPBs and the overall institutional capacities to complete the implementation tasks. Field research in Shanxi also offered the opportunity to interview the provincial EPB head personally to analyse his career background and networks.

A number of constraints for data collection limit the analysis. First, the study only analyses the biographies of 31 EPB heads in office as of 2011 as information on EPB directors in earlier years (i.e. their predecessors) was not available online or was incomplete. Another limitation is that the analysis only examines reported professional affiliations and links, which differ from the full extent of actual links because they do not reflect intertwined informal connections. Some salient information might be also oversimplified, misreported (e.g. formal educational training), or not publically available. In addition, the analysis compared career backgrounds of EPB heads appointed (i.e. the 'successful' candidates) as no data on alternative candidates considered (i.e. the 'unsuccessful' competitors) were available. Additional interviews with the provincial leaders working in organizational departments and Party committees would be necessary to understand why particular EPB heads were picked among the pool of candidates and if and how personal interactions, informal reports, and candidate interviews influence appointment preferences. Moreover, this paper categorizes EPB heads as environmental, business, provincial government, or local government oriented, based on their career path and networks. Of course, an EPB head's orientation is shaped not only by previous work experience and professional ties, but also by a mix of personal motivation, ideological beliefs, values, and career ambitions as well as external circumstances. Additional analysis is also needed to link EPB heads' orientation and actual performance as EPB directors. Despite these shortcomings, to the author's knowledge, this is the first bibliographic database to track the career trajectories of EPB heads and, as such, this study serves as a platform for further research.

The next section introduces concepts and arguments relevant to understanding the career patterns of provincial leading bureaucrats in charge of environmental policy implementation. The paper then analyses the previous work experience, concurrent affiliations, and professional networks of provincial EPB heads, showing that environmental leaders are highly embedded in their locality. The paper concludes by highlighting possible implications of EPB heads' policy orientation and by pointing out possible implications for environmental policy implementation.

Previous Literature and the Assumptions of this Study

This research is based upon a number of assumptions drawn from previous research on the cadre management system and environmental governance in China. These assumptions are as follows: (a) environmental protection law enforcement and policy implementation are flexible and there is ample scope for agency to shape policy outcomes; (b) the EPB head is an authoritative and personally influential figure within the provincial EPB; (c) provincial EPBs operate in a complex system which places them under the local People's Government and the Communist Party Committee and alongside other bureaus; this high level of bureaucratic fragmentation necessitates coordination and consensus-building between these various bodies; and (d) career paths and institutional backgrounds determine the professional skills, networks, factional ties, and incentives of government leaders and bureaucrats. In the following sections, these assumptions are used to address the following questions: (1) What specific work experience, networks, and incentives do Chinese EPB heads have? (2) How do the experiences, networks, and incentives of EPB heads influence the implementation capacity and independence of local EPBs?

Flexible Environmental Law Enforcement and Policy Implementation

Most observers of China are well aware of the gap between stringent national level environmental laws and policies and their sometimes lacklustre implementation at the local level. A number of studies have analysed the implications of implementation flexibility at local levels for the pursuit of environmental goals. For example, a number of studies highlight the existence of 'local protectionism' in China and argue that local governments place more emphasis on protecting their economic interests than on environmental goals (Kostka & Hobbs, 2013; Van Rooij, 2006). While recently there has been progress in creating improved national environmental legislation (Stern, 2011; Xie, 2004) and increasing prioritization of environmental protection work and innovation (Li et al., 2011, p. 132; Wang, 1997, p. 23) aimed at ameliorating the effects of such protectionism, such problems still continue. Some argue that this is because national laws may lack local legitimacy (Bardach & Kagan, 1982; Van Rooij, 2006). Also, bureau leaders have long had substantial discretionary power in how they choose to punish violations of national laws (Davis, 1969), and this still holds true for environmental law. Environmental protection decisions lack internal and external checks and balances (Van Rooij, 2003) and therefore EPB heads, as the leader of the government bureau with primary responsibility for environmental protection,[2] have the opportunity and often the incentive to dilute stringent environmental regulations to satisfy local interests.

Authority and Personal Influence of an EPB Head

Previous studies emphasize the importance of individual leadership at the provincial, municipal, and county levels. Studying the Sichuan provincial EPB, Van Rooij emphasizes that 'the EPB is a centralist institution, and that its leadership has a final (direct and/or indirect) influence on almost all (except for their own positions) major personnel decisions' (2003, p. 46). Another field study in Liaoning attributes major environmental improvements in Benxi municipality to the interventions of a high-profile and charismatic municipal EPB head who 'contrasted sharply with former EPB heads in Benxi, who had held little power' (Bauer, 2006, p. 37). Benxi's powerful EPB head even managed to capture the attention of national leaders, which laid the path for a major local pollution reduction programme (Bauer, 2006, p. 39). These research findings allow us to characterize an EPB head as authoritative and influential within the EPB structure.

Bureau Interdependence and People's Government and Communist Party Committee Authority

EPBs are dependent on the provincial authorities partly because provincial Party and government leaders have the authority over the appointment of provincial EPB heads. The appointment of a provincial EPB director requires the approval of a provincial Party committee as no cadre can be appointed, promoted, or removed without the agreement of a specific Chinese Communist Party (CCP) committee and each appointment is subject to the authorization of the relevant organization departments (Landry, 2008). Thus, provincial Party committees and organization departments directly handle the appointments of EPB heads and, through their influence on personnel decision-making, maintain 'a firm grip on the EPB' (Van Rooij, 2003, p. 46).[3] This is significant because, for provincial leaders, environmental protection is only one target alongside economic growth, employment, and social targets. As such, upper level leaders are likely to appoint a provincial EPB head with views or background who can advance the province's overall development plan. This means that provincial Party and government leaders might often select EPB heads with a background in other areas in order to strengthen EPBs' linkage and ties to other bureaus or to ensure that provincial economic interests are not sacrificed for environmental protection.

In addition to appointment control, there is also a high degree of fiscal and administrative interdependence between the EPB and other government agencies. For EPBs to carry out their daily work and objectives, leading EPB officials depend on support from other departmental heads and local government leaders (Liu et al., 2012). EPBs receive funding for staff and projects from the provincial/municipal government and EPB budgets are thus subject to approval from the provincial/municipal finance bureau (Bauer, 2006; Lo & Tang, 2006). Of course, this gives local governments a high degree of leverage over the work of the EPB. A county-level EPB head in Shanxi complained about the lack of upper level government support:

> For the past ten years, our bureau did not receive any additional staff despite rising environmental regulations and new responsibilities. We have continuously requested new staff trained as environmental specialists but the upper government did not send us any. It is frustrating because without qualified personnel we cannot do our work well.[4]

Another county-level EPB head in Shanxi illustrates the financial dependence:

> In 2011, we could not expand the local central heating area as planned because our municipality faced severe financial pressure this year and the limited financial resources went to other projects. We will apply for national and provincial subsidies but even if we get the money, we need to match it with municipal funding, which is difficult to get at the moment.[5]

Insufficient financial resources also hinder some local EPBs from investing in necessary monitoring equipment. An EPB head in a resource-rich municipality notes:

> We would like to control heavy metal pollution in waste water as we know that some companies pollute it but we cannot afford the monitoring equipment. We have pointed this out to the provincial leaders for many years but only got approval for it this year.[6]

In such an environment of uncertainty, many EPBs therefore need to actively cultivate cooperative relationships and coordinate with other key departments and local leaders. Sinkule and Ortolano's study shows how the Shenzhen EPB employed both formal and informal means to link with other departments. One such formal tactic employed by the EPB, used only in extreme circumstances, is direct appeal to the city mayor to settle a conflict between itself and another department. More commonly used tactics are informal and consist of negotiation, bargaining, and consensus-building in order to influence other departments' actions (Sinkule & Ortolano, 1995, p. 153).

Appointment Decisions, Career Paths, Networks, and Incentives

In addition to developing cooperative relationships with other departments and being locally embedded, the awareness of underlying professional networks is also crucial for understanding EPBs' work and implementation capacities. It is for this reason that gaining an understanding of the career pathways of EPB heads is relevant to the study of localities' implementation capacity. Previous work has examined the career paths of leading government officials and tried to detect patterns in appointment and promotion decisions (Burns, 1987, 1994; Landry, 2008; Manion, 1985, 1993; Shih et al., 2010, 2012; Walder, 1995). While the true basis of personnel decisions remains something of a black box in China, formal criteria considered during appointment processes include qualification requirements—such as age, education, and professional experience—as well as performance evaluation from previous positions, which evaluate cadres' political attitude and moral character. In these performance evaluations, cadre evaluation criteria such as 'both red and expert' (*youhong youzhuan* 又红又专) and 'having both integrity and competence' (*decai jianbei* 德才兼备) are often left purposely vague and are hard to match with operationalizable standards (Mei, 2009). This vagueness leaves ample room for appointing Party and government officials to define what characteristics of a candidate are desirable and which 'merits' are conducive for the organizational good (Tullock, 1965). Appointment procedures are thus shaped by selectors' preferences and career incentives and the selection of particular candidates reveals something about the preferences of 'selectors' (Mei, 2009). As such, the selection of a provincial EPB heads tells us something about the views of provincial Party and government leaders. A deputy head of an Organization Department explains the process through which EPB heads get appointed at the county level:

> First, local county cadres can fill in a democratic recommendation form to suggest suitable candidates for the EPB director position. Next, the [county] Party committee discusses and recommends one or multiple candidates from the recommended list. Afterwards, the [county] organization department picks one or more candidates based on their educational and professional background, administrative and management abilities, and assessment of individual integrity and behavior. We especially pay attention to administrative and management abilities gained during previous work positions because environmental expertise alone does not ensure that the candidate is also a good leader. As a next step, the county government officially announces the nominated candidate(s) and the standing committee of the county people's congress picks the final candidate through a vote. If a single candidate is nominated, a candidate needs to receive more than half of the votes; if multiple candidates are nominated, the candidate with most votes gets selected. Finally, the results will be shared with the municipal organization department. So long as the election was legal, the municipal organization department will approve the candidate.[7]

The above quote highlights the importance of previous work experience in appointment decisions. Provincial leaders can pick between candidates with generalist or specialist/technocratic networks (Shih, 2008). Generalist networks are those not limited to a particular issue area such as environment or economics, while specialized networks are much more limited to these areas. A candidate with generalist networks holds wide-ranging ties and relationships within the Party and government structures, both horizontally and vertically, while a candidate with specialist/technocratic networks holds ties vertically concentrated within a bureaucratic system (xitong 系统) (Shih, 2008).

Many informal selection criteria and 'soft factors' also influence appointment decisions. Li's (2010) empirical analysis of the provincial secretaries and mayors reveals that career promotions based on family ties (largely the domain of princeling, known as *taizi* 太子) or on patron–client relations (personal assistants, known as *mishu* 秘书) continue to be prevalent in China. Other studies have emphasized the importance of factional ties in Chinese politics formed on the basis of being born in the same province (*tongxiang* 同乡), attending the same school (*tongxue* 同学), and sharing common work experience (*tongshi* 同事), with the latter being the focus of this study (see e.g. Hillman, 2010; Li, 1994; Pye, 1995; Shih, 2008; Shih et al., 2010). The networks of an EPB head from previous work experience may shape his or her incentives and policy preferences.

The next section turns to an analysis of the bibliographic data and career trajectories of China's 31 provincial EPB leaders holding office as of February 2011 and shows what kinds of professional networks and links EPB heads hold.

Biographical Characteristics of Current EPB Heads

As has been discussed, EPB heads, to a large degree, determine the relative capacity and roles of local EPBs. Given EPBs' fiscal and administrative dependence on local Party and government organizations, EPB heads' preferences and leadership capacity are determined largely by the quality of their relationship with other actors in the locality. In addition, EPB leaders' network resources and skills are shaped by their previous work experiences. Because the relationships and work experience of EPB heads are determined largely by career path, this section analyses the biographical information of China's 31 EPB heads. It addresses the following question: what specific work experience, networks, and incentives do Chinese EPB heads actually have?

Basic Characteristics

Although basic characteristics did not correlate with other factors such as time in office or career background in statistical analysis, they are provided here as background information on EPB heads' demographics and education. All EPB heads currently serving as EPB director or Party secretary are male. In addition, the vast majority are ethnically Han and between the ages of 46 and 60. The average age is 51.6 years. As shown in Table 1, nearly all EPB heads are nearing retirement, which is mandatory at the age of 60 in China, and are likely to have developed large professional networks over their careers.

From Table 1, we can see that most EPB heads claimed to have 4-year college degrees or higher, and several held Ph.D. degrees. As is often the case for Chinese bureaucrats, many of the EPB heads received degrees or professional training while in the civil service. Their academic fields are diverse, but the majority of EPB leaders studied engineering, economics, and management. Strikingly, just three leaders majored in environmental sciences (EPB head in Tibet) or law (EPB heads in Guizhou and Sichuan).

Figure 1 shows the ages of EPB heads at the time of their appointments to their current positions and their average tenure times in office. The average age

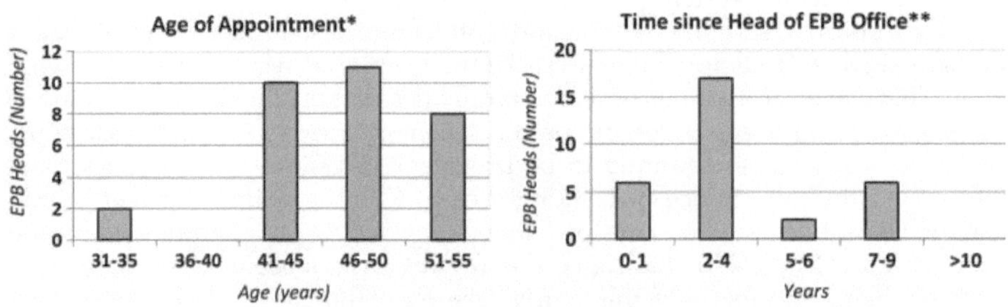

Figure 1. Other characteristics—appointment age and time in office.
Notes: *Average age of appointment: 47.6 years. **Average time in current position as an EPB head is 4.0 years.

Table 1. Basic characteristics and educational backgrounds of EPB heads

	EPB heads (Number)	Percentages (%)
Gender		
Male	31	100
Female	0	0
Ethnicity		
Han	21	84
Minority	4	16
Information not available	7	–
Age range[a]		
56–60	7	23
51–55	11	35
46–50	11	35
41–45	2	6
Total	31	100
Educational level		
Ph.D.	5	18
Master's degree	12	43
4-year college	9	32
2-year college	2	7
Total	28	100
Information not available	3	
Academic majors		
Environment	1	4
Economics/management	9[b]	32
Politics/public administration	2	7
Engineering	9	32
Law	2	7
Chinese	3	11
Others	2	7
Total	28	100
Information not available	3	

[a]Average age of the EPB heads is 51.6 years.
[b]Includes two EPB leaders with degrees in engineering and management.

of appointment for current EPB heads is 47.6. Twenty-three EPB directors (74%) had spent 4 years or less in their position, 2 directors (7%) had spent between 5 and 6 years, while a minority of EPB directors (19%) had served 7–9 years, and none had served for more than 9 years. The average time served in the current position as EPB head is 4.0 years. By comparison, average time served as the head of a provincial Developmental Reform Commission and as the head of a provincial Construction Bureau was 3.56 and 4.60 years, respectively.[8] This is in line with other studies showing that leading Party and government officials serve, on average, 3–4 years before moving on to their next assignment, 1–2 years shorter than the recommended 5-year tenure time for civil servants in leadership positions.[9] EPB directors who have served in their positions for longer than 5 years are the heads of EPBs in Hebei, Shaanxi, Guangdong, Hunan, Yunnan, Tibet, and Chongqing. The longest serving head is the EPB head in Beijing, who took office in July 2002. The heads of EPBs in Hebei, Shaanxi, Guangdong, Hunan, and Tibet were all appointed in 2003, at the beginning of the Hu-Wen administration.

Changing EPB leadership at 4-year intervals affects environmental policy implementation and regulation in numerous ways. Short tenure cycles can result in leading officials prioritizing short-term projects that deliver tangible results within their own tenure over projects with a longer time frame (Eaton & Kostka, 2012). Therefore, newly appointed EPB heads might have a strong interest in selecting environmental projects that will bear fruit quickly and tacitly sideline policies with a longer time to maturity. For instance, EPB leaders may opt to temporarily shut pollution-intensive companies to fulfil annual sulphur dioxide targets while putting off the more difficult tasks of either closing down the business permanently or developing long-term solutions, such as offering training or devising incentive mechanisms for technological advancements. High turnover in EPB leadership can also frustrate environmental policy implementation in more prosaic ways since newly appointed EPB heads, the majority of whom are not familiar with environmental regulations and policies, must spend considerable time in getting to know the lay of the land, which can result in damaging discontinuity in environmental initiatives (Eaton & Kostka, 2012). Finally, short tenures of EPB directors also offer provincial leaders more frequent chances to appoint replacements, thereby enhancing the political control of organization departments and Party committees (Huang, 2002; Landry, 2008).

Career Paths and Institutional Backgrounds

This section summarizes the previous positions and career backgrounds of EPB heads. Table 2 lists the positions that the current EPB heads held immediately prior to their appointments. From this table, we can see that only eight EPB heads were promoted within the EPB, while the remaining three-fourths came from positions outside the environment field. Almost one-third of the current EPB heads were appointed from municipal level positions such as municipal mayor, deputy mayor, Party secretary, or vice Party secretary.

Table 3 categorizes the career backgrounds of EPB heads. The EPB heads were coded according to the government level or type of bureau in which they spent most of their careers. Twenty-eight out of 31 EPB heads spent from 10 to 20 years on a particular career path. Three EPB heads are not easily categorized because their careers did not follow a 'standard' career path.[10] For example, the Tibet EPB head began his career as a researcher and an aid worker in the environmental field before joining the civil service as the deputy head of the Tibet EPB. From Table 3, we can see that 16 EPB heads (52%) spent most of their career at

Table 2. Position prior to the appointment as the EPB head

	EPB heads (Number)	Percentages (%)
Within the EPB	8	26
Within another provincial department	7	23
Economic	3/7	
Other	4/7	
PPC/Party committee	4	13
Municipal mayor (deputy)/Party secretary (vice)	10	32
SOE	1	3
County-level Party secretary	1	3
Total	31	100

Table 3. Career background

	EPB heads (Number)	Percentages (%)
Career backgrounds[a]		
EPB	7	23
Provincial level	16	52
Government/Communist Youth League	*3/16*	
Economic bureau	*8/16*	
Other bureau	*6/16*	
Municipal level	5	16
Enterprise	2	6
Total	31	100

[a]Refers to the majority of time of EPB heads' career spent in the civil service.

the provincial level, 7 within the EPB (23%), 5 in lower level governments (16%), and 2 with a state-owned enterprise (SOE) (6%). The majority of EPB heads built their careers in political positions at the provincial governments or Party Committees, provincial non-economic departments, or municipal governments. Next to long political careers, many EPB heads spent their careers in economic or business positions, either in an economic bureau career or in an enterprise career.

EPB leaders who rose through the EPB ordinarily began as environmental protection researchers or pollution control specialists. They were promoted through the bureau ranks until reaching the position of EPB head. Leaders with careers in provincial economic bureaus started out as cadres in provincial economic bureaus and were rotated among the different provincial government economic agencies. EPB heads who originated from municipal governments spent nearly all of their careers in a single locality, gradually climbing the bureaucratic ladder. EPB heads with prior careers in non-economic bureaus ordinarily spent most of their careers rising through the ranks in a single bureau before eventually being rotated through leadership positions in other provincial non-economic bureaus. Two of three leaders who worked in provincial governments or Party committees were promoted through the Communist Youth League system. There are two SOE career EPB heads who began as workers in industrial enterprises and rose to upper management positions in an SOE before being appointed to positions in the government late in their careers.

Interestingly, EPB leaders with environmental and other non-economic backgrounds are most commonly appointed to urban municipalities. All four province-level municipal governments appointed EPB leaders with careers in environmental protection or non-economic bureaus such as labour and social security, urban planning, or construction.

Figure 2 analyses EPB heads' previous work experience in SOEs. Nineteen per cent of all EPB heads (or 6 out of 31) worked at some point for an SOE. Three EPB heads worked for enterprises in heavy industry (iron and steel, metallurgy, and chemicals), which could be especially useful work experience for an EPB head. For example, the Jiangxi EPB head worked his entire career in a provincial chemical SOE, beginning as a factory assistant and eventually rising to the chairman of the board. The Gansu EPB director also spent most of his career in an enterprise, rising from a technician to a deputy CEO. The high share of EPB heads with previous work experience in SOEs is partly the result of organizational departments' control over personnel appointments across SOEs and government and Party organizations, making cross-postings easier and more widespread.

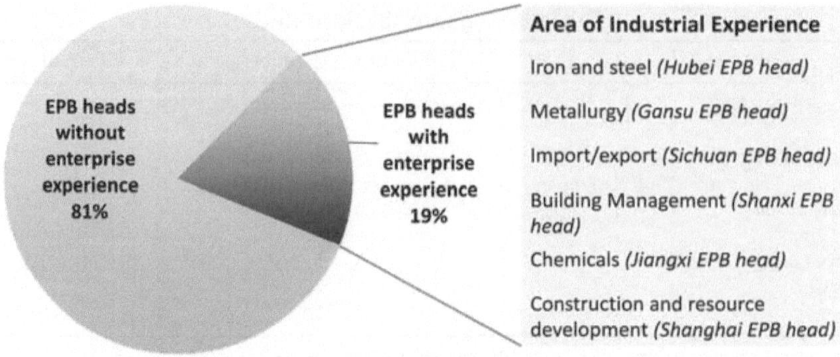

Figure 2. EPB heads' enterprise experiences* (*100% = 31 heads*).
Note: *Experience refers to temporary placements, concurrent affiliations, previous jobs, or rotating positions in SOEs.

Concurrent Affiliations and Cross-postings

Of the 31 provincial EPB profiles analysed here, seven EPB heads list outside affiliations and cross-postings on their CVs (Table 4). Three EPB heads hold National People's Congress (NPC) delegate positions (Sichuan, Shanghai, and Anhui), which are very prestigious titles, and two EPB heads hold Provincial People's Congress (PPC) delegate positions (Guizhou and Hebei). One EPB head is cross-posted as the Head of the Nuclear Safety Management Bureau (Sichuan). Two EPB heads have affiliations with local universities, one as a Ph.D. adviser (Shandong) and the other as a professor (Hunan). The number of affiliations of EPB heads is likely to be understated. For example, it is common for a bureau director to serve as a PPC delegate, perhaps often enough that many EPB heads do not list it in their short biographies published online. In addition, EPB heads may not publicly reveal their affiliations with local enterprises to avoid the appearance of bias.

How to Interpret Previous Work Experiences and Career Backgrounds

This section addresses the following question: how do the experiences, networks, and incentives of EPB heads influence the implementation capacity and indepen-

Table 4. EPB head enterprise experience.

	EPB Heads (Number)	Percentages (%)
EPB Leaders with Enterprise Experience[a]		
Yes	6	19
No	25	81
Total	31	100
EPB Leader	*Areas of Industrial Experience*	
Hubei	Iron and steel	
Zhejiang	Metallurgy	
Sichuan	Import/export	
Shanxi	Building management	
Jiangxi	Chemicals	
Shanghai	Construction and re source development (affiliation)	

dence of provincial EPBs? As the *guanxi* and network ties of an EPB head from his or her previous work experience may shape the incentives of EPB heads, these links and networks can be used to assign 'types' to each of the provincial EPB heads. Each type can be inferred to come with a possible orientation or slant towards the interests of his or her corresponding networks and factional ties.

As has been stated previously, these 'types' are based on the career backgrounds of EPB heads, most of whom spent most of their careers working in a particular area. Table 3 categorizes EPB directors by the institutions in which they spent most of their careers. From this table, it is clear that most EPB heads fall into one of four career background categories and are nearly evenly split among them: careers in EPBs; in provincial governments, Party committees, or non-economic provincial bureaus; in local governments as municipal mayors or Party secretaries; or in economic bureaus or SOEs. Here these career backgrounds are considered to be of the following types: 'environmentally oriented' for career EPB bureaucrats; 'provincially oriented' for those bureaucrats who spent most of their careers in provincial governments, Party committees, or non-economic provincial bureaus; 'locally oriented' for bureaucrats with careers in local government; and 'business oriented' for career economic bureaucrats and SOE managers. Figure 3 presents the different career background categories.

If existing networks based on career backgrounds influence EPB implementation capacities and priorities, then the selection of a provincial EPB head with a certain background can have a significant impact on local environmental protection. Since EPB heads can be relatively easily categorized by their work backgrounds, we can infer that provincial leaders seem to prefer particular types of politically or economically embedded officials when considering bureaucrats for promotion to EPB head. In the appointment process, provincial leaders can be expected to weigh the benefits of EPB head candidates' professional networks, skills, and incentives. Given the limited available information on the candidates under consideration, provincial leaders will likely consider a candidate's profile carefully and initially use a candidate's career background, institutional affilia-

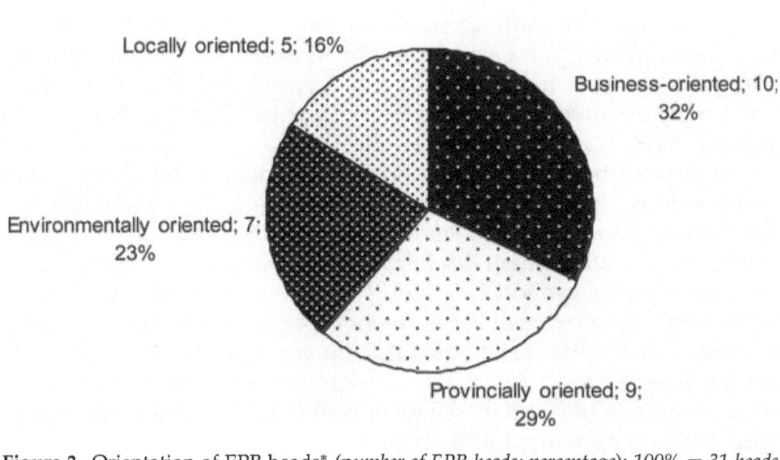

Figure 3. Orientation of EPB heads* (*number of EPB heads; percentage*); *100% = 31 heads.*
Notes: * Categorization is based on the analysis of three previous work experiences. Policy-setting/provincially oriented category combines leaders with government and non-economic provincial bureau careers. The business oriented category combines leaders with careers in provincial economic bureaus and enterprises.

tions, and previous performance evaluations as an indicator for their preferences and abilities. Career paths can also convey a great deal of information to the appointing officials on the candidate's professional networks, factional ties, skills, and incentives. As has been argued below, the 'best' candidate may not, in all cases, be the person with the greatest ability and credentials, but rather be the person who best supports the political position of the appointer. The following section introduces the characteristics of these EPB head orientation categories and explains how each category of background may influence EPB implementation capacities and priorities in different ways.

Environmentally Oriented EPB Heads

Environmentally oriented leaders are departmental heads who were promoted entirely within the EPB structure. Leaders belonging to this category have developed ties to other local EPBs and actors closely linked to EPB-related work. Over the years, these government officials have accumulated in-depth environmental knowledge and expertise at the sub-provincial and provincial levels. Their technical and environmental expertise combined with an awareness of local implementation practices may make them effective leaders in tackling difficult environmental challenges, but only if they are given sufficient support from the provincial government and other bureaus. Having worked most of their careers as environmental specialists, drafting or carrying out national and provincial environmental regulations, these leaders are likely to have large networks in the environmental protection field and yet may lack extensive connections with other non-environmental actors. This relative lack of networks beyond the EPB may hinder their ability to capitalize on their specialized knowledge on environmental protection methods as they may be less effective in establishing and coordinating working relations with other departments.

Provincially Oriented EPB Heads

Provincially oriented EPB heads spend the majority of their careers at the Provincial People's Government or Party offices or in non-economic bureaus such as urban planning, construction, or labour and social security. The interests of these provincially oriented EPB leaders are likely to be aligned with the province's policy-making level. Their behaviours will be the most in keeping with the policy objectives and priorities of the provincial government and Party committee because of previous work experience which provides the networks and factional ties to the province. In contrast to environmentally oriented cadres, provincially oriented EPB heads with career backgrounds at the provincial level are more likely to have a wider range of networks with other provincial departments. Bureaucrats from the provincial government or Party committee will have overseen the work of many provincial bureaus. Non-economic bureau leaders are regularly rotated among provincial level bureaus once they have reached the peak of their careers and often have accrued a great deal of inter-bureau work experience by the time they are appointed EPB heads.

However, leaders in this group who spent nearly their entire career at the provincial level may have limited awareness of local implementation difficulties and practical concerns. Also, leaders with provincially oriented careers in non-economic bureaus may lack strong networks with economic bureaus or enterprises.

They may, thus, have less practical knowledge of business and fewer incentives to support provincial business interests except when there is an overall welfare benefit to the provincial government such as tax revenue. From a professional skills angle, on the other hand, certain non-economic bureau heads may have increased implementation capacity because they share some characteristics with environmentally oriented leaders. Given their backgrounds in urban planning, construction, and labour, leaders in this category could effectively help provincial leaders to meet particular cadre targets. For example, an EPB leader with a bureau background in construction or urban planning is likely to be especially adept at enforcing national guidelines and devising provincial standards to lower energy usage in the local construction industry or addressing rising pollution in large urbanized cities. Yet, this may be very dependent on the level of provincial support for these policies.

Locally Oriented EPB Heads

Locally oriented leaders refer to EPB heads who spent most of their career working in county and municipal governments and agencies. Many officials in this cohort of EPB leaders worked most of their career in one or two localities before being promoted to the post of provincial EPB head. They are likely to retain tight links with their home localities. These ties likely provide a platform for informal feedback, which may improve provincial implementation polices. Of course, such cosy relationships might also result in favouritism and exception granting. Lacking necessary networks and influence at the provincial level, locally oriented leaders might also be less effective in gaining buy-ins from other provincial players and enterprises.

Beyond these considerations, during their many years spent at the local level, these leaders likely develop a political sense of how to balance competing priorities in the policy implementation process. This is particularly relevant to environmental protection. The ambitious approach to environmental protection in national and provincial policies and the pragmatic economic and social concerns of government agents and business interests at the local level make the implementation of these policies especially contentious. The awareness of local implementation practices and knowledge about existing policy loopholes gives these leaders an advantage when enforcing environmental mandates that are highly dependent on municipal support.

Business Oriented EPB Heads

Finally, the EPB leaders categorized as business oriented have worked previously in provincial economic bureaus or state-owned companies. An economic bureau background ensures that a leader is well situated in the provincial policy-setting, while an enterprise background signals familiarity with implementation practices at the enterprise level. The business oriented 'type' can draw on wide range of economic and business-related contacts in forming business–government alliances or breaking down enterprise opposition to stringent regulations. Existing networks with other economic bureaus and enterprises likely extend the reach of the EPB, thus possibly strengthening it. However, highly business-oriented leaders might also sacrifice strictly enforcing environmental protection laws and policies in favour of provincial economic growth or individual business interests. For example, while working in an SOE, an EPB leader might have inherited business contacts, potentially giving rise to a conflict of interest. Also, simply having work experience in an SOE does not imply that a leader understands or has contacts with other enterprises. Having worked predominantly at the provincial level on economic issues, leaders in this category may also lack expertise in environmental matters and local policy implementation.

In summary, there are different types of EPB heads based on their integration into provincial politics (policy level *vs.* local level) and existing networks based on career backgrounds (generalist *vs.* specialized networks). Delineating different types of EPB leaders helps to understand provincial EPB leadership and the implementation capacities of provincial EPBs, along with the appointment preferences of provincial leaders.

Discussion: Appointment Decisions and Typology Preferences of Provincial Leaders

The above documentation of the diversity of types among EPB heads is surprising. EPB leaders' career backgrounds may differ vastly, ranging from environmental bureaucrat, SOE business manager, political strategist, to implementers. What can explain these differences among appointed EPB leaders? This begs the following questions: what are provincial governments or Party leaders considering when appointing a local EPB head? Do appointment patterns tell us anything about the needs of a province as perceived by the provincial government or Party? Or do they reflect leaders' career and political incentives?

A number of studies have highlighted the CCP's ability to select government and Party leaders in a rational manner (Burns, 1987, 1994; Huang, 2002; Manion, 1985; Walder, 1995). The findings of these studies indicate that the selection of bureau leaders is rarely random or haphazard. Candidates are selected through a rigorous evaluation process which considers a wide number of factors, such as past work experience and performance. The 'best' candidate is not necessarily the candidate who will most thoroughly enforce environmental regulation laws or even achieve the most progress in environmental protection. Provincial leaders and organization departments, when appointing a provincial EPB director, consider both overall provincial preferences and personal incentives. From the possible typologies, provincial leaders will need to select one 'type'. The most desirable candidate for the EPB head position will likely be an official whose assumed networks and incentives suggest to provincial leaders that, under the new leadership, the EPB's practices will be favourably aligned with provincial priorities and with parochial interests of the provincial Party secretary and mayor. The implications of these preferred backgrounds are examined in the remainder of this section.

Provincial Priorities and Cadres' Career Incentives

When assessing bureaucrats for promotion to EPB head, provincial leaders will consider the overall provincial context and weigh potential trade-offs, balancing considerations of the need for economic development and the complexity of environmental pollution problems. Provincial governments have different economic and environmental priorities, depending on the province's level of economic development (rich/poor), the level and complexity of environmental pollution (high/low), and the required level and type of implementation (locally/provincially and coordinated/uncoordinated). Provincial leaders may therefore seek out different types of environmental leaderships and appoint a candidate who balances a certain mix of environmental or business expertise, provincial or local networks, and high or low career ambitions. These factors may heavily influence the capacity of an EPB head and consequently strengthen or weaken the environmental protection institution itself.

In addition to considerations related to the overarching provincial interest, personal preferences and career incentives of provincial Party secretaries and mayors are at least equally important, if not more so. A number of personal incentives may influence appointment decisions, including potential of appointees to affect a leader's ability to meet targets, receive annual bonuses, and be promoted, along with personal favouritism, but the most important of these is the desire for promotion. Rationally, cadres will evaluate candidates based on whether they can help them to advance, or at least safeguard, their own careers.

For example, because Party secretaries and mayors have signed individual Cadre Responsibility Contracts, outlining economic, social, environmental, and political targets and goals necessary for their annual evaluation, candidates who will help them meet these goals, particularly those of mandatory or 'hard' (*ying xing* 硬性) targets, are likely to be found on the top of the candidate list. For environmental policy implementation, provincial leaders have clearly defined environmental protection targets, some of which are mandatory such as energy intensity reduction targets. Among the many different targets, provincial leaders may have identified some that are more difficult to meet. If a particular problem is significant, provincial leaders are likely to appoint an ambitious candidate who will go out of his or her way to ensure (mandatory) that targets are met. For example, if it is challenging to meet provincial energy efficiency targets, provincial leaders may opt for an EPB head with the strength or expertise to turn the wheel around. If, by contrast, economic targets are comparatively harder to meet, a business oriented EPB leader might be more appropriate. Adjustments in cadre evaluation criteria, or signals from Beijing of impending adjustments, along with changes in career incentives and the overall political climate, further affect the EPB selection preferences.

Examples of appointed EPB heads in Shanxi and Inner Mongolia show how selecting a business oriented or locally oriented candidate can help to address a particular provincial concern and/or can assist provincial leaders in meeting particular cadre evaluation targets.

Selection of a Business Oriented EPB Head: Shanxi

Shanxi faced increasing national and international media coverage for its severe air pollution in 2005 and 2006,[11] which put air pollution issues on the front burner for provincial leaders. In 2006, the provincial leaders picked a 'strong' and highly authoritative candidate with a large business-related network able to take drastic steps and deal with media pressure. The candidate selected by the province, who remains the head of Shanxi Province's EPB as of March 2012, started his political career as the Deputy Secretary of the First Business Bureau in Taiyuan City in 1985 and later worked as Party secretary of Shanxi Province Supply and Marketing Cooperative Council. Prior to joining the civil service, Shanxi's EPB head was in the military and also worked as a television editor and journalist. He claimed to use his military experience to get 'respect' from other leaders and environmental protection workers.[12] As for the media experience, within the Shanxi EPB, he was known for being particularly adept at handling media attention and obtaining media coverage for the province's significant environmental progress since his appointment in 2006.[13] Over the last 6 years in office, this 'strong' EPB head made substantial progress in reducing emissions, particularly SO_2 emissions, and enforcing overall environmental protection standards, partly because he had a clear mandate from the provincial government following the National Development and Reform Commission's introduction of hard targets in emissions reduction and energy savings for provincial leaders. One of his first actions as a newly appointed EPB director was to temporarily revoke some county EPBs' right to conduct environmental evaluations (*quyu xianpi* 区域限批) to punish them for turning a blind eye to coal and coking operations without approved environmental impact reports.[14] This drastic measure effectively blocked counties' ability to approve new industrial projects. This was the first time that a provincial EPB director had made use of this severe penalty and the ensuing media glare put the affected county leaders under extreme pressure (Eaton & Kostka, 2012) and was 'an effective warning to other counties'.[15] The EPB head also devised new incentive mechanisms to motivate local EPB heads by, for example, promising a financial award from the provincial EPB to the municipality that could first be removed from the list of the nation's most polluted cities. Yangquan municipality was the first to clear its name, in 2008, and was awarded 2 million RMB by the province.[16]

Selection of a Locally Oriented EPB Head: Inner Mongolia

While provinces such as Shanxi, which was under external pressure to improve air pollution, may select a candidate with the ability and credentials to effectively implement environmental mandates, the still-developing regions may appoint a candidate who can balance economic growth and environmental protection concern. Inner Mongolia, for instance, selected a candidate who lacked career incentives and large business-related or provincially oriented networks that facilitate the forming of bridges between different government agencies. The head of the Inner Mongolia EPB as of February 2011 belongs to the Mongol ethnic minority, a contrast to the majority of departmental heads in Inner Mongolia who are ethnically Han. He has a 2-year college degree in a non-environmental science field and was promoted to the office close to retirement at the age of 55, much later than the average appointment age of 47.6 years. Prior to appointment as the EPB head in 2006, he worked in the Inner Mongolia Provincial Economic and Trade Commission and later as a mayor in Baotou municipality, one of Inner Mongolia's fastest growing and most resource-rich municipalities. In contrast to the EPB head in Shanxi, he held no previous non-governmental work experience. Given his previous work experience on economic issues, he might be a candidate who gives way when pressured by the Inner Mongolian provincial leaders to prioritize commercial interests over more strict environmental enforcement.

Reported environmental data of the previous years confirm poor provincial environmental performance. Between 2005 and 2009, per capita CO_2 emissions rose faster in Inner Mongolia than in any other province in China and by 2009 Inner Mongolia had the highest per capita CO_2 emissions among all provinces (i.e. 12.85 metric tons of CO_2 per capita) (Du et al., 2012, p. 374). In September

2010, 3 months before the end of the 11th Five-Year Plan evaluation period, leading provincial officials in Inner Mongolia further indicated during interviews that the province was far from meeting its energy saving and emission targets.[17] Provincial leaders saw environmental protection and economic growth as a contradiction (*maodun* 矛盾) and did not want to miss out on provincial growth that had picked up rapidly since 2006 due to large incoming foreign investments. Even officials working within the EPB repeatedly stated that the province could not prioritize energy saving and emission reduction targets at the expense of economic development. A local EPB official summarized this prevailing attitude as follows: 'we are doomed to "pollute first, manage later" [*xian wuran, hou zhili 先污染后治理*] because we need to accumulate enough money and technology before we can afford to protect the environment'.[18] Without strong leadership in the EPB and support from provincial leaders, environmental policies were implemented only halfheartedly. An official in charge of pollution control shared his perception that the provincial EPB had little enforcement power:

> In 2009 we started to move polluting enterprises in Hohhot beyond the 2nd ring road and outside of Hohhot. But the progress is very slow. There is very little we can do as an EPB because the EPB only plays a supervisory and reporting role in the process. We do not have the authority to order electricity or water companies to cut off utilities to these enterprises. If we would do so, we would get sued.[19]

By contrast, in Shanxi, the provincial EPB head widely advertised the 'cut electricity, cut water' (*duandian duanshui* 断电断水) policy that could be implemented because the provincial EPB head had successfully coordinated with local state-owned electricity and water companies to cut off a company's access to utilities.

Candidate Selection and Selection Preferences: Provincial Level Municipalities

In contrast to a somewhat prevalent government–enterprise coordination approach to policy implementation in provinces and autonomous regions, there is evidence that urban provincial level municipal leaders in Beijing, Shanghai, Tianjin, and Chongqing prefer EPB candidates with environmental expertise or specialized networks with other non-economic bureaus. All four centrally administered municipalities appointed EPB leaders with careers in environmental protection or non-economic bureaus such as labour and social security, urban planning, or construction. In addition, EPB heads in all four municipalities are young, with leaders in Beijing, Shanghai, and Chongqing being appointed to their positions at the age of 45 or less. Since younger EPB heads are likely to have smaller networks and EPB heads with environmental protection or non-economic bureau backgrounds will have backgrounds which complement EPB work, this paper suggests that provincial level municipal leaders prioritize environmental expertise and networks with other provincial level departments over a business network-based implementation capacity of EPBs. Provincial level municipal leaders may also prefer a relatively independent, professionally skilled EPB head.

Conclusion

This paper has examined career backgrounds and institutional backgrounds of local government officials in provincial EPBs in China. EPB leaders' career backgrounds differ vastly, ranging from environmental bureaucrat, SOE business manager, political strategist, to implementers, despite EPB heads assuming seemingly similar environmental protection and law enforcement tasks. Based on their career backgrounds and their associated networks, EPB leaders fall into the following four typologies, and are nearly evenly split among them: environmentally oriented, business oriented, policy-setting/provincially oriented, and implementation/locally oriented. The analysis of EPB heads' career backgrounds shows that only one-fourth of the provincial EPB heads were promoted within the EPB, while the remaining three-fourths came from positions outside the environment field. These findings have implications for inferring the unique characteristics of a province's EPB leadership and the implementation capacities of provincial EPBs.

The distinct characteristics of these different 'types' of EPB leaders show that provincial leaders make important appointment decisions among clearly differentiated candidates. This may have implications for inferring the appointment preferences of provincial leaders. Provincial mayors and Party secretaries and leaders in the provincial Organization Department, when appointing EPB heads, balance different overall provincial and personal preferences. Certain candidate 'type' and career background might be more suitable to meet these different preferences. When assessing bureaucrats for promotion to EPB head, provincial leaders will consider the overall provincial context and weigh potential trade-offs, balancing considerations on the need for economic development, the complexity of environmental pollution, and required implementation practices. Rationally, provincial leaders select an EPB head who will act in the best interest of the province as a whole as perceived by the leaders while also helping provincial leaders to bolster their own political careers.

This paper further argued that a leader's local networks and relationships likely improve the implementation capacity of EPBs, but also at times weaken incentives to strictly enforce environmental protection laws and policies. This implies that there is a delicate balance between networks which advance implementation capacity and those which paralyse environmental protection efforts in China. EPB heads who are well connected to other key departments and enterprises are more likely to be successful in creating cross-sectoral and cross-departmental coalitions. However, the professional networks which may improve environmental protection capacity in a province through coordination and consensus-building also further decrease the already limited autonomy of local EPBs. Highly embedded leaders with limited independence may place emphasis on commercial, economic, or political interests over environmental protection and turn a blind eye to polluting enterprises or illicit practices of other government bureaus.

Acknowledgements

The author thanks William Hobbs for his excellent research assistance and Sarah Eaton, Thomas Johnson, Arthur Mol, and Ran Ran for their valuable feedback on the earlier drafts of this paper.

Notes

1. EPB 'head' refers to the EPB Director (*juzhang* 局长) or the EPB Party secretary (*dangzu shuji* 党组书记), which are positions almost always held concurrently.

2. The main role of the EPB is to enforce environmental protection laws by regulating and monitoring pollution in water bodies, air, noise, solid waste, and soil. For a more detailed description of the role of EPBs, see Lee (2006, p. 82).
3. A comprehensive review of cadre appointment processes is beyond the scope of this paper. For further details on cadre appointments and the role of the Chinese Communist Party's organizational departments, see Manion (1985, 1993), Burns (1987, 1994), and McGregor (2010, pp. 70–103).
4. Interview, September 2011, Shanxi, county EPB head.
5. Interview, September 2011, Shanxi, county EPB head.
6. Interview, September 2011, Shanxi, municipality EPB head.
7. Interview, September 2011, Shanxi, deputy head of county Organization Department.
8. The source here is the author's database consisting of biographic information for 124 departmental heads of provincial EPBs, development and reform commissions, economic commissions, and construction bureaus in China's 31 provinces.
9. For example, Persson and Zhuravskaya (2012, p. 56) show that the average tenure on the job for 755 provincial Party secretaries during 1980 and 2005 was 3.29 years. Kostka and Yu (2012) collected data for 898 Party secretaries at the municipal level and found that while in the 1990s the average tenure time was 4.2 years, the average tenure time dropped to 3.3 years between 2002 and 2011. Landry's work on 2058 municipal mayors in office during 1990–2001 also suggests that tenure times steadily declined from an average of 3.2 years in the 1990s to 2.5 years by 2001 (2008, p. 90).
10. The three exceptions to the patterns of local political or economic careers are the EPB heads in Jiangsu, Anhui, and Tibet. The Tibet EPB head, the youngest provincial EPB leader, began his government career as the deputy head of the Tibet EPB only three years before being appointed as the EPB head. He was 29 when he left Beijing to do aid work in Tibet in 1998 after spending most of his twenties studying for his Ph.D. in environmental protection and after working briefly at the Beijing Environment Science Institute. His career background was categorized as 'EPB'. In Anhui, the head of the provincial EPB spent a nearly equal amount of time, over ten years each, in both provincial economic departments and in lower level government positions, but spent slightly more time in provincial economic departments. He was also categorized as 'economic'. The Jiangsu EPB head held for 6 years the position of a vice secretary (*fu mishu* 副秘书) in the provincial government, but served most of his earlier career in different economic government bureaus. He was categorized as 'economic'.
11. See, for example, *China Daily* (17 July 2006) and Watts and Rameesh (2007).
12. Interview, July 2010, Shanxi, provincial EPB head.
13. Interview, July 2010, Shanxi, government official at the provincial EPB.
14. Interview, July 2010, Shanxi, provincial EPB head.
15. Interview, July 2010, Shanxi, provincial EPB head.
16. Interview, July 2010, Shanxi, provincial EPB head.
17. Interview, September 2010, Inner Mongolia, government officials at the provincial Economic Commission and EPB.
18. Interview, September 2010, Inner Mongolia, county EPB head.
19. Interview, September 2010, Inner Mongolia, government official in the Pollution Control Office.

References

Bardach, E. & Kagan, R. A. (1982) *Going by the Book: The Problem of Regulatory Unreasonableness* (Philadelphia, PA: Temple University Press).

Bauer, J. R. (2006) *Forging Environmentalism: Justice, Livelihood, and Contested Environments* (New York: M.E. Sharpe).

Burns, J. (1987) China's Nomenklatura System, *Problems of Communism*, 36(5), pp. 36–51.

Burns, J. P. (1994) Strengthening Central CCP Control of Leadership Selection: The 1990 Nomenklatura, *China Quarterly*, 138, pp. 458–91.

China Daily (2006) Rivers Run Black in Shanxi Province, July 17. Available at http://www.china.org.cn/english/environment/174874.htm (accessed 4 January 2013).

Davis, K. C. (1969) *Discretionary Justice: A Preliminary Enquiry* (Urbana, IL: University of Illinois Press).

Du, L., Wei, C. & Cai, S. (2012) Economic development and carbon dioxide emissions in China: Provincial panel data analysis, *China Economic Review*, 23, pp. 371–384.

Eaton, S. & Kostka, G. (2012) Does Cadre Turnover Help or Hinder China's Green Rise? Evidence from Shanxi Province, Frankfurt School Working Paper Series No. 184, Frankfurt School of Finance and Management, January 2012.

Hillman, B. (2010) Factions and spoils: Examining political behavior within the local state in China, *China Journal*, 64, pp. 1–18.

Huang, Y. (2002) Managing Chinese bureaucrats: An institutional economics perspective, *Political Studies*, 50, pp. 61–79.

Kostka, G. & Yu, X. (2012) Local Leadership in China: Career Mobility and Career Advancement Opportunities, Frankfurt School Working Paper Series, Frankfurt School of Finance and Management, September 2012.

Kostka, G. & Hobbs, W. (2013) Embedded Interests and the Managerial Local State: The Political Economy of Methanol Fuel-Switching in China, *Journal of Contemporary China*, 22(80), pp. 1–15.

Landry, P. (2008) *Decentralized Authoritarianism in China: The Communist Party's Control of Local Elites in the Post-Mao Era* (Cambridge, NY: Cambridge University Press).

Lee, S. (2006) *Water and Development in China: The Political Economy of Shanghai Water Policy* (London: World Scientific Publishing).

Li, C. (1994) University networks and the rise of Qinghua graduates in China's leadership, *Australian Journal of Chinese Affairs*, 32, pp. 1–30.

Li, C. (2010) China's midterm jockeying: Gearing up for 2012—Part 1–3, *China Leadership Monitor*, vols. 31–33, pp. 1–24 (Part 1), pp. 1–24 (Part 2), pp. 1–19 (Part 3).

Li, Y. W., Miao, B. & Lang, G. (2011) The local environmental state in China: A study of countylevel cities in Suzhou, *The China Quarterly*, 105, pp. 115–132.

Liu, L., Bing, Z. & Bi, J. (2012) Reforming China's multi-level environmental governance: Lessons from the 11th Five-Year Plan, *Environmental Science & Policy*, 21, pp. 106–111.

Lo, C. W. H. & Tang, S. Y. (2006) Institutional reform, economic changes, and local environmental management in China: The case of Guangdong province, *Environmental Politics*, 15, pp. 190–210.

Ma, X. & Ortolano, L. (2000) *Environmental Regulation in China: Institutions, Enforcement, and Compliance* (Oxford: Rowman & Littlefield Publishers).

Manion, M. (1985) The cadre management system, post-Mao: The appointment, promotion, transfer and removal of party and state leaders, *China Quarterly*, 102, pp. 203–233.

Manion, M. (1993) *Retirement of Revolutionaries in China: Public Policies, Social Norms, Private Interests* (Princeton, NJ: Princeton University Press).

McGregor, R. (2010) *The Party—The Secret World of China's Communist Rulers* (London: Penguin Books).

Mei, C. (2009) Brings the politics back in: Political incentive and policy distortion in China, Doctoral Thesis, University of Maryland, 2009.

Persson, P. & Zhuravskaya, E. (2012) Elite Influence as a Substitute for Local Democracy: Evidence from Backgrounds of Chinese Provincial Leaders, Working Paper SSRN, Version August 2012.

Pye, L. W. (1995) Factions and the Politics of Guanxi: Paradoxes in Chinese Administrative and Political Behaviour, *The China Journal*, 34, pp. 35–53.

Shih, V. (2008) *Factions and Finance in China: Elite Conflict and Inflation* (New York: Cambridge University Press).

Shih, V., Shan, W. & Liu, M. (2010) Gauging the elite political equilibrium in the CCP: A quantitative approach using biographical data, *The China Quarterly*, 201, pp. 79–103.

Shih, V., Adolph, A. & Liu, M. (2012) Getting ahead in the communist party: Explaining the advancement of central committee members in China, *American Political Science Review*, 106(1), pp. 166–187.

Sinkule, B. J. & Ortolano, L. (1995) *Implementing Environmental Policy in China* (Westport: Praeger Publishers).

Stern, R. (2011) From dispute to decision: Suing polluters in China, *The China Quarterly*, 206, pp. 294–312.

Tullock, G. (1965) *The Politics of Bureaucracy* (Washington, DC: Public Affairs Press).

Van Rooij, B. (2003) Organization and procedure in environmental law enforcement: Sichuan in comparative perspective, *China Information*, 17(2), pp. 36–64.

Van Rooij, B. (2006) Implementation of Chinese environmental law: Regular enforcement and political campaigns, *Development and Change*, 37(1), pp. 57–74.

Walder, A. G. (1995) Local governments as industrial firms: An organizational analysis of China's transition economy, *American Journal of Sociology*, 101(2), pp. 263–301.

Wang, Y. (1997) Quanmian Jiaqiang Huanjing Fazhi Jianshe Quebao Kua Shiji Huanbao Mubiao De Shixian [Overall Strengthening the Establishment of the Environmental Legal System in Order to Guarantee the Realization of the EP Goals When Moving into the Next Century], *Environmental Work News Report*, vol. 12, pp. 16–24.

Watts, J. & Rameesh, R. (2007) Fighting for air: Frontline of war on global warming. *The Guardian*, March 26. Available at http://www.guardian.co.uk/environment/2007/mar/26/globalwarming.china#article_continue (accessed 4 January 2013).

Xie, Z. (2004) Shuli He Luoshi Kexue Fazhanguan Tuidong Huanbao Gongzuo Zai Xin Taijie [Establishing and Implementing a Scientific Development Doctrine to Promote Environmental Work onto a New Level], *Huanbao Gongzuo Ziliao Xuan [Selected Materials on Environmental Protection Work]*, 4(4), pp. 4–14.

Political Modernization in China's Forest Governance? Payment Schemes for Forest Ecological Services in Liaoning

DAN LIANG* & ARTHUR P.J. MOL**

**China National Forestry Economics and Development Research Center, State Forestry Administration, No. 18, He Ping Li Dong JieBeijing, 100714, People's Republic of China*
***Environmental Policy Group, Wageningen University, Hollandseweg 1, 6706 KN, Wageningen, The Netherlands*

ABSTRACT *Payment for environmental services (PES) schemes are increasingly being introduced in developed and developing countries for the ecological conservation of forests also. Such payment schemes resemble a new mode of forest governance labelled political modernization, in which centralized and state-based command-and-control policies make room for market dynamics, non-state actors, and decentralization. In entering the new Millennium, China has massively started using payment schemes to conserve its forests. An analysis of the implementation of the Forest Ecological Benefit Compensation Fund Programme in Liaoning Province is used to investigate whether China's PES schemes resemble notions of political modernization. It is concluded that Liaoning Province introduced market dynamics and farmer participation in the implementation of its PES scheme, but in a way different from that theorized by political modernization scholars. Hence, it should rather be seen as a 'Chinese style' political modernization process.*

1. Introduction

During the second half of the twentieth century (until at least the late 1980s), China exploited its forests for fuelling its economic boom (Liang, 2012). Deforestation for wood harvesting, agricultural development, and energy development came together with ineffective forest protection policies. Not only did the forest stocking volume decrease, but also the ecological quality of the forests deteriorated. But three developments have given China's forest policies a very different look since the turn of the Millennium. Most importantly, the devastating flood along the Yangtze River in 1998 sounded an alarm on the enduring environmental degradation and depletion of forests in the country. The flooding proved that the regulation and enforcement of forest management in the past decades had failed to meet the objective of providing a sustainable forest ecosystem. Since 1998, China has launched six national forest ecological conservation and restoration programmes to combat forest destruction and to reduce negative impacts on forest ecosystems in China.[1] This was possible because the increasing fiscal revenue following rapid economic development and the 1994 Fiscal Reform strengthened the central government's financial ability to provide pubic goods, including ecological restoration and environmental protection. Finally, during the last decade, the forest tenure reform decentralized collectively owned forest-land, distributed it over to individual farmer households, and, with that, allowed forest resources to be governed through individual financial incentives. As such, over the past decade, forest ecosystem protection has changed from mainly state-controlled and regulated forest harvesting and production to forest protection and ecological restoration using a variety of new (financial) instruments and approaches and involving a diversity of actors. This seems to resemble western ideas and models of the so-called political modernization in European forest governance, where decentralization and participation, a strong focus on ecological services next to wood production, and the use of economic approaches have emerged (Arts & Buizer, 2009; Kotilainen *et al.*, 2008; Veenman *et al.*, 2009).

This article looks into some of the innovations of forest governance in China by investigating the establishment of a public benefit forest (PBF) and the financial Forest Ecological Benefit Compensation Fund Programme (FEBCFP), against the background of ideas and practices of political modernization as articulated in European forest governance. PBF, established around 2000, is a category of forests where no commercial harvesting is allowed. The simultaneously established FEBCFP compensates forest owners for subsequent income loss and has an annual national budget of up to 3 billion Yuan (US$428.6 million). In addition, this national programme has succeeded in involving provincial and local governments to develop their own payment schemes for forest ecosystem services. In 2009, 25 provinces had established their own provincial payment schemes for forest ecological services, including the densely forested north-eastern province of Liaoning. Few empirical studies have assessed the functioning, implementation, and performance of these forest payment schemes, in terms of their contribution to ecological services, how these market-based approaches work in China's forest governance, and to what extent local farmers actually participate in the design and implementation of these schemes. Hence by taking the FECFP in Liaoning Province as a case, we look into three aspects of China's payment schemes for forest ecological services: the ecological effects of these payment schemes, the consequences of these schemes for local livelihoods, and the degree of participation of local actors in the schemes.

After framing payment for environmental services (PES) schemes in terms of political modernization (Section 2), we provide an overview of forest policy and payment schemes in Liaoning Province. Subsequently, a methodology for assessing payment schemes is introduced (Section 4). In Sections 5–7, we report on a detailed analysis of the local implementation of payment schemes in Benxi and Xinbin counties, Liaoning Province, and finish with conclusions.

2. PES as Political Modernization

Following widespread criticism of the achievements of conventional command-and-control environmental policies in the 1970s and 1980s (Jänicke, 1986), the

1990s witnessed a variety of experiments and innovations in new policy-making and implementation arrangements. In the European literature, these governance innovations have been brought together under the label of political modernization: the modernization of environmental politics (Blowers, 2002; Jänicke, 1993; Mol, 2002, 2003; van Tatenhove & Leroy, 2003; van Tatenhove *et al.*, 2000); in the US literature, a more or less equivalent process of policy innovation has been identified, often labelled reinventing government (Rosenbaum, 2000).

In the European literature, political modernization refers to a set of interdependent governance innovations that together form an alternative governance model for the conventional command-and-control mode of governance. While not restricted to the field of environmental protection and nature conservation, most of the elaborations on political modernization draw their empirical material from these sectors. The modernization of politics and policy is then brought in line with the ideas and theories of ecological modernization, also for China (China Centre for Modernization Research, 2007; Mol, 2006; Mol *et al.*, 2009; Zhang *et al.*, 2007). The decentralization of environmental policy-making, the use of new (financial and voluntary) policy instruments and approaches, and the involvement and participation of wider non-state constituencies in policy-making and implementation are some of the key characteristics of political modernization. Political modernization scholars identify these new tendencies in contemporary environmental governance, but some also claim that such innovations make environmental governance better adapted to the new social constellation of our time and hence more effective in addressing current ecological challenges than conventional command-and-control modes of governance.

China's mainstream environmental policy-making and implementation system is still strongly based on a centralized state that applies a command-and-control regulatory strategy to design and implement policies and plans top down. But this mainstream model is not static. Quite a few scholars have reported on experiments and (niche) innovations in China's environmental governance institutions (Carter & Mol, 2006; Chen, 2009). Besides a remarkable growth in the capacity of the environmental state, in government environmental investments, in environmental laws and regulations, and in environmental monitoring, various authors have also referred to changes that seem to resemble some of the political modernization tendencies identified in Western Europe. The decentralization of policy-making capacity (Jiang, 2006), the further use of market and economic approaches (Zhang & Wen, 2008), and the sprouts of public participation (Zhong & Mol, 2008; various contributions to this special issue) are at least the subjects of experimentation in China. China's payment schemes for environmental services seem to bring a number of such policy innovations together.

But one can expect that political modernization developments as they have unfolded in environmental governance of western democracies over the past three decades will only be partially similar to the environmental policy innovations currently visible in China. Although similarities can be found in the failures of a centralized command-and-control system of environmental governance that trigger governance innovations, the context of a strong centralized state, with little transparency, a very particular market system, and little room for bottom-up civil society, may colour political modernization in China differently. Hence, in applying the concept of political modernization to analyse environmental policy innovations in China, recognition should be given to these differences, rather than applying a strict—European—interpretation of political modernization.

The concluding section will revert to the question of the applicability of the concept of political modernization to contemporary China.

2.1 *Payment for Environmental Services*

The on-going and increasing degradation of ecosystems around the globe has resulted in a search for new approaches to preserve ecosystems and the ecological services that they provide. PES, as one of the new measures, is increasingly being employed in developed and developing countries as an innovative tool in ecological protection and conservation by bridging the gap between the supply and the demand of environmental services and channelling financial flows from beneficiaries to service providers.

PES is a voluntary transaction where a well-defined environmental service or a land use likely to secure that service is being 'bought' by a service buyer from a service provider under the condition that the service provider secures service provision (Wunder *et al.*, 2008).

PES moves away from conventional state-dominated ecological conservation and environmental protection programmes, by involving new non-state actors, new market-based dynamics, and a decentralized approach. As such, PES should be interpreted as part of a political modernization process. In the forest sector, PES has been applied to various environmental services including watershed services, biodiversity, soil erosion prevention, and carbon sequestration at different scales. Regardless of its wide application, the performance of PES has not yet been carefully examined. Costa Rica, famous for its PES initiatives, has arguably witnessed most evaluative studies (Miranda *et al.*, 2003; Sánchez-Azofeifa *et al.*, 2007; Wunder *et al.*, 2008; Zbinden & Lee, 2005).

However, only recently has the information on the growing number of PES schemes introduced in China been disclosed with few details on the implementation and performance of these PES schemes (see Bennett, 2009; Zheng & Zhang, 2006; Zhang *et al.*, 2008). Forest conservation policy is arguably the main sector where payment schemes for environmental services are being applied in China.

3. Liaoning Province and Forest Policy

Northeast China is a region abundant with forest resources and is regarded as one of the important timber production bases in the country. Liaoning Province, in the southern part of China's northeast, consists of 14 municipalities and 74 counties and districts (Figure 1). With 4641 thousand hectares, its forest coverage is 31.84%. However, the forest resources are unevenly distributed over the province. Almost 70% of its forest area, 82.6% of the forest stocking volume, and 90% of the natural forests are concentrated in the eastern part and most of the functions of natural forests such as watershed conservation can also be found in this region. Under a continental monsoon climate, Liaoning Province is fragile and frequently experiences drought and floods. Soil erosion and other natural disaster events (e.g. major stock loss) are worsened by uneven precipitation and a mountainous terrain prone to erosion, but even more by anthropogenic factors, including farming on sloping land, harvesting for firewood, overstocking of wild silkworms (tussah), and mining (Wang, 2008; Zhang *et al.*, 2005). The frequency of floods and droughts

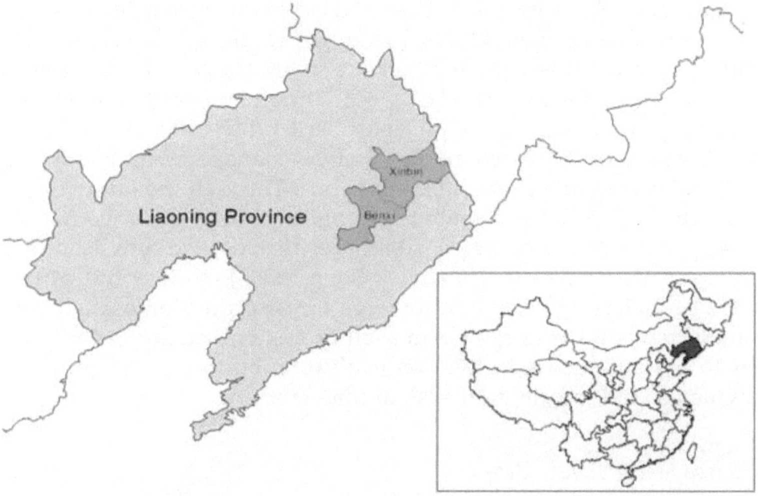

Figure 1. Liaoning Province in China and the location of the case-study counties.

in the eastern part of Liaoning Province has changed dramatically since 1980. Furthermore, soil erosion has extended even to the more forested eastern part.

3.1 *Forest Policy*

Facing serious environmental degradation, the central government radically changed its forestry policy, moving the focus from mainly timber production to afforestation and ecological conservation. In 1984, China's first forest law introduced forest management strategies based on a five-category classification system of forest resources: protection forest, timber forest, economic forest, fuel forest, and special-use forest (Liang, 2012). The division of the categories was controversial since any forest might provide multiple services. However, it created the possibility of restricting the discretion of local forest authority on forest use, which usually preferred harvesting forests for timber. But due to the power exercised by local governments in classifying forests, the majority of forests in China were still categorized as timber forests (Dai *et al.*, 2009). In 1988, China's forest law was amended to reclassify the five forest categories into two: commodity forest and PBF for ecological purpose (Dai *et al.*, 2009). This two-category forest classification system juxtaposed ecological functions with commodity purposes to further emphasize the importance of ecological attributes of forests. However, there was little incentive for local forest authorities to enlarge PBF area for ecological purposes. Moreover, after the 1994 fiscal reform, the central government established its own tax collection system and seized a majority share of the taxes. Local governments suffered from shrinking resources for supporting PBFs. Between 1996 and 2000, the central forest administration struggled and bargained with other national ministries to channel funding for PBFs. Only in 2001 were large amounts of resources directed from the finance ministry to set up payment schemes for forest ecosystem services in 11 provinces, including Liaoning.

Since then, the central government has issued a series of regulations to direct the management of PBFs, such as *Measures for Demarcating and Validating National*

PBF and *Management Measures for the Central Compensation Fund for Forest Ecological Benefit*. Following these national rules, Liaoning Province set up its own management institutions for PBFs—*Liaoning Province's Detailed Rules for Implementing the Management of Public Ecological Benefit Forest* (in brief: *Detailed Rules*). The Detailed Rules (Liaoning Provincial Forestry Department (PFD), 2006a) stipulate formal rules for classification, protection, productive management, monitoring and inspection, and punishment for non-compliance. Through the target responsibility system, which includes agreements or contracts between governments with a set of targets, a point system, responsibility certificates, and provisions for monitoring of the performance (Guttman & Song, 2007), the central government directed the provincial government towards forest protection and conservation. For PBF management, the target responsibility system usually includes multiple targets (with measurable indicators) on institution-building, PBF protection, the use of payments, and information system management.

3.2 *Collective Forest Tenure Reform*

The forest tenure reform is relevant for understanding the implementation of payment schemes in Liaoning. Liaoning Province started its collective forest tenure reform in March 2005, 3 years before the national policy was officially publicized in the 2008 *Guidelines on Fully Promoting Collective Forest Tenure System Reform*. In Liaoning, all villages participated in the forest tenure reform (Xu *et al.*, 2008). In this tenure reform, the use rights of collective forestland are distributed to individual farmer households, delimited with clear boundaries and put under protection of legally binding certificates.[2] The objective of the reform is to turn forest resources into production factors which can be exchanged and deployed in a market-based economy and to further improve the efficiency of forestry production.

Furthermore, the tenure reform also aims to establish a new relationship between property rights and forest conservation. Under the former system of forestry property rights, the government tried to regulate forest management of the villages directly through village committees. As a result, the farmers were excluded from forest management and were only regarded as a threat to forest resources. However, simply cutting the connection between the forest and the people around it, who usually have more knowledge on and access to local forest resources, actually stimulates more illegal harvesting. Therefore, the new system of forestry property rights is set up to include local people in forest management and conservation.

Historically, most of the forests in the east of Liaoning Province were used for timber harvesting and fuel for daily use. After the demarcation of PBFs, the forests have changed from an economic resource to an environmental asset. The payment from the schemes (see below) can only partly compensate the loss of income to local farmers and communities.

There is a major controversy among forestry authorities whether PBFs should be included in the forest tenure reform (State Forestry Administration, 2006) and the central government should allow each province to decide whether or not the tenure reform includes PBFs. In the past, the responsibility of managing and protecting PBFs resided with villages (often with the village committees) and forestry bureaus only need to make a responsibility contract with each village (see Figure 2). Forestry bureaus hired foresters to protect the PBF from fire, pest,

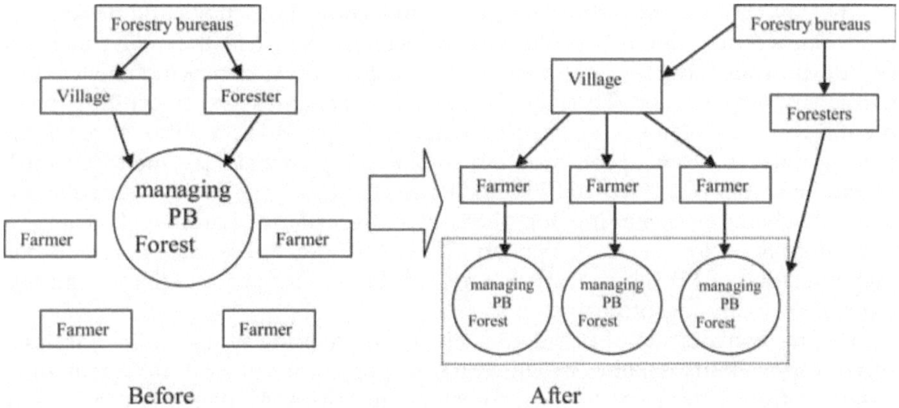

Figure 2. Change in PBF management due to the tenure reform.

disease, and illegal harvesting; local farmers had no access to the PBF and no responsibility to manage the PBF. But forestry bureaus have to face hundreds and thousands of farmer households if the PBF is handed out to farmers. In addition, under low payment standards, a PBF will be a burden for its owners rather than an income source.

Liaoning PFD extended the tenure reform to PBFs, in which local farmers shared responsibility for ecological conservation of PBFs (see Figure 2). The decentralization of the 'ownership' of PBFs facilitates participation among farmers, who are then treated more as cooperative partners for protection than as potential threats to the forests. At the same time, with the tenure reform responsibilities, obligations and rights were redefined and redistributed among governments, villages, and farmers (Figure 2). Individual farmers obtained rights to use their plots of PBFs to develop agroforestry and other production as long as they did not hinder ecological services, but they also had to share the responsibility of PBF protection through a management contract with village committees. Depending on their willingness and the size of their family plots,[3] farmers can delegate PBF protection to foresters or perform it themselves. Through the tenure reform, local farmers moved from a relatively marginal position to the centre of PBF use and management.

The tenure reform provides an institutional space for bringing forestland into economic circulation, although detailed regulation for that is still missing. However, it is much harder for the owners of PBFs to apply for secured loans from banks than for those of commercial forests. The forest tenure reform enlarged the gap between the profit to be earned through commercial forests and the payment for PBFs.

3.3 *Payment Schemes*

The payment schemes are meant to support PBFs. In Liaoning Province, three types of PBFs exist: national PBF, provincial PBF, and provincial natural forest. The definition of a PBF is determined by forestry departments at the provincial and county levels.

The payment schemes stipulate a series of technical functions and basic rules of procedure concerning PBF protection. In Liaoning Province, the main functions of a PBF are water source conservation, soil erosion prevention, wind shelter, and biodiversity conservation. Three types of payment schemes to support the protection of forest ecosystems have been initiated in the province: a national payment scheme,[4] a natural forest protection scheme,[5] and a provincial payment scheme.[6] In total, 2.65 million hectares of PBFs and natural forests have been included in the payment schemes, accounting for 41.8% of its forestland. Liaoning Province—much more than, for instance, Yunnan (Chia, 2010)—gives room for some economic utilization of PBFs by permitting non-timber forestry production, including seed collection and agroforestry.

The implementation of payment schemes in Liaoning relies on the administrative responsibility system, in which the responsibility of PBF management is consigned from provincial to local forestry authorities. Management tasks and responsibilities are divided among county governments, forestry bureaus, township forest stations, village committees, and forest rangers (see Figure 3). Township forest stations employ forest rangers to manage and protect PBFs and county forestry bureaus provide regular training and examination for them. The payment is distributed and used through financial and forestry departments, according to a set of management measures.

The redistribution of forests to farmers caused a problem regarding the legitimacy of the payment schemes since the PBF contracts were initially signed between the country forestry bureaus and the village committee. But after the tenure reform, the *de facto* owners of PBFs are local farmers. Therefore, new contracts had to be introduced between farmers and the forestry authorities. It became imperative to clarify the obligations and rights for PBF owners. In 2009,

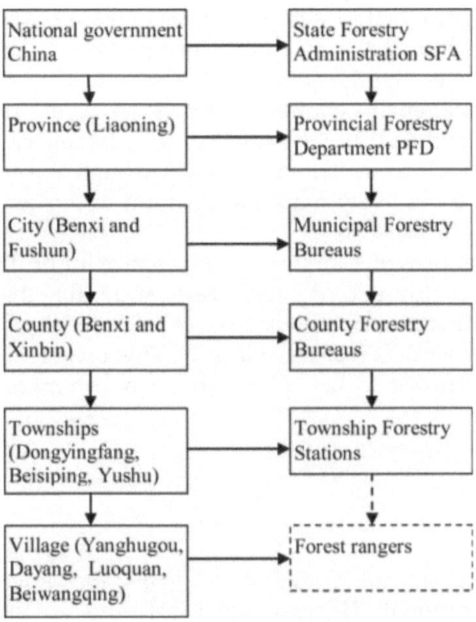

Figure 3. China's state forest administrative system (with case studies between brackets).
Note: forest rangers are non-state actors.

the Liaoning PFD (2009) issued the *Opinion on Facilitating Reform on the Management and Protection Mechanism on PBF* and formally started a province-wide reform on the PBF management. A 2-year plan was made to settle the responsibility of PBFs by various contracts with local forest owners.[7] This plan also aimed to channel a larger part of the payment to individual farmer households directly. This part of the payment includes replanting and tendering subsidy (22.5 Yuan per hectare) and the additional part of the new payment policy (75 Yuan per hectare; 10 Yuan per mu) since 2009 (it was 5 Yuan per mu before 2009). The execution of this plan is still on-going, and according to our 2010 survey, only one-third of the farmer households have made management contracts with the county forestry bureaus.

4. Methodology for the Assessment of Payment Schemes

4.1 *Research Design*

In assessing payment schemes for forest services, we applied a case-study approach, focusing on both process and outcomes. Hence, we look at the outcomes of the schemes in terms of forest conservation and changes in farmers' livelihood practices, as well as at the participation of local farmers and local authorities in the design and implementation of PES schemes.

The measurement of ecological performance—the most important objective of payment programmes—is difficult, especially for forest protection programmes, for three reasons. First, there exists a major time lag between protection activities and ecological outcomes. Second, besides payment schemes, many intervening factors may influence the outcome, which cannot be ruled out easily. Finally, it is difficult to acquire adequate data on ecological quality, especially in developing countries such as China. In this study, a rough assessment was made of the ecological effect of PES, using two data sources: observations of local farmers measured through household questionnaires and forestry resource inventories on the quality of the forest ecosystem. The impact of payment schemes on the livelihood practices of farmer households was analysed in this study through household questionnaires and interviews with officials, focusing on income changes of farmer households; changes in firewood collection; impacts on livestock agriculture and cash cropping in forests; and effects on other income sources for communities. Participation focuses on how and how much targeted recipients have participated in the formulation and implementation of the payment schemes. Household questionnaires, interviews with officials, and farmer focus groups were used to collect data on the actual participation at the different stages of the payment programmes.

4.2 *Site Selection*

To empirically investigate the implementation of payment schemes, two case-study areas in Liaoning Province were selected, taking into account the following criteria: the starting time of the implementation of the payment schemes (the earlier, the better), the area of PBFs, the presence of forests with high ecological and economic importance, and the existence of the forest tenure reform. Benxi County and Xinbin County[8] proved to be excellent case-study areas, fulfilling these criteria the best. Benxi County and Xinbin County (see Table 1), respectively,

Table 1. Situation of sample counties in 2010

County	Township	Area (km^2)	Population (1000)	Forest coverage (%)	Area PBF (km^2) National	Provincial[a]
Benxi	14	3344	300	73	1203	943
Xinbin	15	4287	310	73	1153	1234

Source: county forestry bureaus.
[a]Provincial PBF includes natural forest under the Liaoning Province Natural Forest Protection Programme.

under the administration of Benxi City and Fushun City, lie in the main forested region of Liaoning Province along the Taizi River, which is the primary branch of the Liao River. Most PBFs of the counties are used for conserving water sources and preventing soil erosion. Since 2001, these two counties have been included in the pilot of the central fiscal payment scheme. Since 2005, the forest tenure reform has been implemented in these counties, as one of the first trials of tenure reform on PBFs nationwide. In addition, Benxi County and Xinbin County have sizeable Chinese traditional medicine planting and mushroom and fungi cultivation.

Four villages were selected from the two counties using sampling criteria similar to those used for the counties (Table 2). All four villages are located in mountainous areas but have good access (with cement roads) to towns and cities. Yanghugou Village ranks first in area, population size, forest resources, and PBFs. Beiwangqing Village has a socio-economic profile similar to that of Yanghugou, but has few forestlands and PBFs. Dayang and Luoquan are quite similar in area, forestlands, and PBFs, and both have a relatively high income *per capita* than the other villages.

4.3 *Data Collection Methods*

To understand the general implementation of payment schemes and collective forest tenure reform in Liaoning, in-depth, semi-structured interviews were conducted in 2010 with four officials from the Ecological Benefit Forest Management Office and the Forest Tenure Reform Office in Liaoning PFD. Interviews were also carried out with five officials of municipalities and county forestry bureaus to understand the implementation of payment schemes at the county level to collect related documents and to select sample villages.[9]

After the interviews, a survey of farmer households in the four villages of Benxi County and Xinbin County was conducted. Between 12 and 15 farmer households were randomly sampled in each village, from a list provided by the village leader. A total of 54 farmer household heads (all male) were interviewed on income and livelihood practices, forest resources and environmental improvements, participation in payment schemes and forest tenure reform, and attitudes towards the payment schemes. Two sample counties were minority regions and 83% of the respondents were Manchu. All respondents completed primary education and 85% finished middle school. Over 60% of the interviewed households had four to six members and 89% of the households had less than two children (aged below 18 years).

Table 2. Socio-economic situation of sample villages in 2010

County	Township	Village	Area (km^2)	Household	Population	Cropland (ha)	Income *per capita* (Yuan)	Forestland (ha)	PBF (ha)	
									National	Provincial
Benxi	Dongyingfang	Yanghugou	58	378	1397	168	4400	5467	4400	367
Benxi	Dongyingfang	Dayang	24	297	1075	153	6500	2133	800	400
Xinbin	Beisiping	Beiwangqing	52	409	1380	375	4000	4520	546	1763
Xinbin	Yushu	Luoquan	28	168	621	38	5000	2580	433	700

Source: interviews with local village leaders.

5. Assessing the Ecological Performance of PES

The first assessment of PES is related to its core objective: to what extent did the payment schemes deliver in terms of improved ecological performance? Forest inventories and the farmer survey were applied to assess the ecological performance of PES schemes in Liaoning.

No village-level quantitative data of PBF stocks were available to assess the change in PBF resources after the introduction of the payment schemes. Hence, provincial level data were used. Interviews with provincial and lower level forest administrators gave no reason to expect that ecological developments in PBFs in our case-study villages significantly deviate from the provincial averages. Overall, the PBFs in Liaoning Province improved consistently and even faster in recent years (Table 3). Under the payment schemes, forest coverage in PBFs increased from 77% in 2005 to 79% in 2008 (see also Xu *et al.*, 2008, p. 12). In addition, the growth of stock volume accelerated at a rate of 5.4% in 2008 (compared with only 1.5% in 2005). The extension of forest area and increase in stock volume caused a decline in the age structure of PBFs in which young and middle trees form the main part (92% in 2008). Forests with a young age structure are more vulnerable to extreme climate and to forest pests and diseases. The farmer household survey showed that this was caused by the forest tenure reform and the afforestation incentives in PES for local farmers: 19% of the respondents had recently afforested PBF land as the government provided a subsidy of 50 Yuan per mu for afforestation.

According to governmental reports (Table 4), the main threats to PBFs include forest fires, pests and diseases, and illegal harvesting and occupation. While forest fire has been controlled quite successfully, a large area of forests is still affected by pests and disease every year. The young age structure and single tree species made the PBFs vulnerable to pests and diseases. In addition, the number of forestry administrative punishment cases (a rough indicator of illegal harvesting) significantly decreased after the 2005 forest tenure reform. By redistributing PBFs to individual farmer households, the reform created incentives for local farmers to manage and protect PBFs. The farmers were encouraged to use PBFs to develop agroforestry and they became more active in protecting PBFs. However, local economic development widened the gap in revenue between PBFs and industrial development. It motivated illegal occupation of PBFs for mining and construction activities.

In addition to these provincial level forest inventories, village-level farmer surveys were used to support the assessment of ecological improvements of the payment schemes. Almost half of the farmers believed that their villages had no apparent soil erosion before the payment schemes were implemented. About

Table 3. PBF changes in Liaoning Province

Year	Coverage	Annual growth in stock volume	Age structure[a]
2005	77%	1.4%	40:32:11:14:3
2008	79%	5.4%	44:48:4:3:1

Source: Liaoning PFD (2006b, 2009).

[a]The age structure shows the area proportion of stands at different ages: young, middle, pre-mature, mature, and over-mature, respectively.

Table 4. Main threats to PBFs in Liaoning Province

Year	Total area PBF (10^3 ha)	Forest fire-affected area (ha)	Forest pest and disease: Affected area (ha)	Forest pest and disease: Percentage treated	Illegal occupation (ha)	Law cases[a]
2005	2650[b]	61	277,530	86%	0	593
2008	2650[b]	81	255,351	95%	13	197

Source: Liaoning PFD (2006b, 2009).
[a]The law cases refer to forestry administrative punishment cases related to PBFs; most of these are illegal harvesting.
[b]In both years, Liaoning PBF measured 2,290,000 ha of the so-called important PBF.

33% of the respondents thought that their village had some soil erosion and 20% thought that widespread soil erosion existed before the schemes were implemented. In Dayang Village and Beiwangqing Village, most of the farmers thought that their village had some soil erosion. The respondents in Luoquan Village indicated no apparent soil erosion before the implementation of the schemes, while about half of the respondents of Yanghugou Village thought that the village had widespread soil erosion. Annual precipitation in Yanghugou Village is over 800 mm, more than that in the other sample villages (740 mm), and 80% of the precipitation occurs in July and August. In addition, steep sloping land extensively exists in Yanghugou Village. Therefore, among the case-study villages, it is most vulnerable to soil erosion.

Around 24% of the respondents thought that soil erosion had been significantly reduced since the introduction of the schemes, 46% of them thought that some reduction had taken place, and 30% observed no change in soil erosion following the payment scheme implementation. Almost all the respondents of Luoquan Village, where no apparent soil erosion had existed previously, thought that the schemes brought no change to the local environment. In contrast, almost all the respondents of Yanghugou Village (where widespread soil erosion existed) thought that soil erosion has been reduced slightly. In Dayang Village and Beiwangqing Village, most of the respondents thought that soil erosion was reduced. Although perceptions of local farmers remain interpretations and should not be used as exact or absolute measurements of ecological situations, they are valuable in indicating change in ecological services over time. Together with the forest inventories, these farmer household data lead to a cautious conclusion that the payment schemes did bring about relative ecological improvements in terms of reduced erosion in ecologically fragile areas and increased forest coverage and growth in stocking volume in PBFs.

6. Assessing the Impacts on Livelihood Practices

Since the forest tenure reform had been launched just after the implementation of the payment schemes and the effects of the reform and the payment schemes intertwined with each other, it is not always easy to quantitatively separate the impacts of payment schemes on the livelihood practices of farmer households from the impacts of the tenure reform. Therefore, we analysed the combined impacts of the tenure reform and the payment schemes.

Before the implementation of the forest tenure reform, the PBFs in the sample villages were directly managed by the villages (Figure 2). The payment for PBF management and protection was given to villages, and villages selected—with the approval of county forestry bureaus—forest rangers (and team leaders[10]) and paid them from the subsidy. Although the forest rangers in principle should be village members, the amount of payment was so little that the payment did not have substantial impacts on the overall livelihood of the villages (Table 5). Only those who were employed as forest rangers benefited from the payment schemes. This distribution of payments strengthened the protection of PBFs, but local farmers mostly did not profit.

Moreover, stringent implementation of PBF protection resulted in a negative impact on the livelihood of local farmer households. Half of the farmers thought that the payment schemes negatively impacted their income; 35% asserted that the payment schemes had not affected their income, and a few farmers (7%) saw their income increasing with the schemes.[11] Negative impacts were especially experienced in reduced fuel wood harvesting (57% of the households) and limitations on animal husbandry in forests (41%). Respondents indicating negative income effects by the schemes were not evenly distributed among the villages, with 79% in Luoquan, 62% in Yanghugou, 47% in Dayang, and 25% in Beiwangqing. Luoquan is a typical village dependent on forestry production with the largest forestland area *per capita* (4.15 ha) and the smallest cropland area *per capita* (0.06 ha) among the four villages. In addition, its forest quality is highest with a stocking volume ranging from 105 to 120 m^3/ha. Therefore, farmers in Luoquan Village economically suffered most from the forest harvesting ban of the payment schemes. Likewise, Yanghugou Village has a similar natural resource structure for local livelihood, and 87% of its forestland has been demarcated as a PBF, which restricted logging activities. As a result, the payment schemes had greater impacts on Luoquan and Yanghugou than on Dayang and Beiwangqing.

The 2005/2006 collective forest tenure reform had a more significant impact on the local economy than the payment schemes. It reshaped the economic structure, contributed to the redistribution of forest resources, improved the forestry investing environment, and developed agroforestry. According to the survey, the average household income increased by 75% from 11,144 Yuan in 2001 to 19,528 Yuan in 2009 (Table 6). All sources of income increased at least nominally after the implementation of the payment schemes and the reform, but timber harvesting and agroforestry resulted in a dramatic rise in income. The rapid rise in timber income was the result of the redistribution of commercial forest as the revenues now flowed directly to local farmers. The reform also triggered a substantial

Table 5. Forest rangers and total payment for forest services to each village and annual average between 2001 and 2009

Village	Population	PBF (ha)	Forest rangers	Payment (Yuan/ year)	Annual payment per ranger
Yanghugou	1397	4767	12	115,200	9600
Dayang	1075	1200	6	57,600	9600
Beiwangqing	1380	2309	4	38,400	9600
Luoquan	621	1133	3	28,800	9600

Source: interview with local village leaders.

Table 6. Average household income structure and change, 2000 and 2009 (in Yuan and not corrected for inflation)

	2000		2009		% increase/decrease	
Income source	Yuan	%	Yuan	%	Absolute	Share
Cropping	2887	26%	3719	19%	29%	−7%
Animal husbandry	506	4.5%	898	4.5%	78%	0%
Off-farm work	3892	35%	6057	31%	56%	−4%
Timber harvesting	37	0%	2328	12%	6185%	12%
Agroforestry	1006	9%	2509	13%	150%	4%
Small business and others	2817	25.5%	4017	20.5%	43%	−5%
Total	11,144	100%	19,528	100%	75%	–

Source: farmer household survey.

increase in agroforestry—mainly cultivating Ginseng, wild vegetables, and Korean pine in PBFs. The reform adopted a new paradigm to manage PBFs, which aimed at easing conflicts between ecological conservation and local economic development. As the 10 Yuan per mu payment from the central and provincial governments is still considered quite low (see Figure 6 below), this new paradigm offers an additional incentive for protecting PBFs. In Liaoning Province, agroforestry has evolved into an institutionalized practice to complement insufficient governmental payment for PBFs.

Under a substantial increase of the average household income, its structure has also changed. On average, off-farm work was still the main source of household income in 2009 (31%), but its share decreased somewhat and so did the share of agricultural production in average household income. Compared with the decline in traditional agriculture and off-farm work, forestry-related income experienced a rapid rise and contributed significantly to household income (Table 6).

7. Participation in Forest Governance

To what extent were farmers involved in setting up and implementing the new market-based forms of forest conservation and restoration? This section reports on how farmers participated in the collective tenure reform and in policy formulation, demarcation, management, and examination of the payment schemes.

7.1 *Participation in Collective Tenure Reform on PBFs*

The collective forest tenure reform as a comprehensive institutional change includes different components and implementing stages: distributing forest plots to farmers (clarifying boundaries of forest plots, contracting out plots, and issuing certificates of forest tenure), establishing supporting policies (forest tenure exchange market, forest tenure for loan policy, forest insurance policy, disclosure policy on forest harvesting information, forestry association or cooperation policy, etc.), and a reform of PBF management and protection. We focus on the distribution of forest plots to farmers, which is the principal part of the reform.

The central and local governments have emphasized the participation of local farmers during the reform (Liaoning Provincial Government, 2005). Initially, each village established a working team for the reform, which thoroughly investigated forest resources of the village and the willingness of the villagers to distribute collective forestland. The core part of the reform consisted of making an implementation arrangement on how to distribute collective forestland and to divide corresponding rights and obligations. At least two-thirds of the villagers or village representatives had to agree with the draft implementation arrangement. After the villagers accepted the arrangement, it was sent to the township government for approval. Then, the village working team confirmed and registered tenure, boundaries, and areas of each parcel of forestland and made a contract with each farmer household according to the arrangement. The inventory and contracts were sent to the township governments for checking. Finally, the municipal and county forestry bureaus examined the performance of the reform.

According to our survey, 94% of the respondents were familiar with the forest tenure reform arrangement of the villages and all of these were satisfied with the final tenure reform, mainly because it clarified and legalized forest tenure. In addition, 98% of the respondents supported the redistribution of the tenure of PBFs from villages to farmer households, because the reform facilitated payment distribution,[12] improved PBF protection, and/or enhanced possibilities for the development of agroforestry (Figure 4). After the implementation of the reform, 92% of the respondents received formal forest tenure certificates issued by county forestry bureaus.

7.2 *Participation in Policy Formulation of PES*

Current payment policies have been developed at the national and provincial levels. The process of policy formulation followed a traditional government-directed policy-making paradigm. The payment schemes in Liaoning Province apparently have been shaped by two forces: the central government and the so-called Two Meetings of the Provincial People's Congress (PPC) and the Provincial People's Political Consultative Conference (PPPCC).[13] The central government initiated the central payment scheme and provided both funding resources and an institutional base for the provincial payment schemes. Local representatives had some influence through the 'Two Meetings' in Liaoning Province. For example, in 2008, 12 PPC representatives promoted a proposal to the Meeting on establishing payment schemes for water conservation area around the Dahuofang reservoir. This proposal was ranked with the highest priority for approval

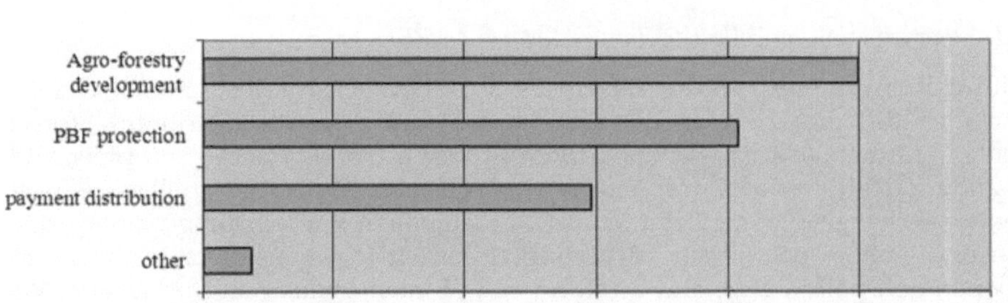

Figure 4. Reasons for supporting PBF tenure redistribution ($n = 54$).

and implementation. In 2008, the provincial financial department earmarked 150 million Yuan for paying ecological services in the eastern region—the major water conservation area, which also includes Xinbin County.

But there is little evidence on the direct involvement of local farmers in payment policy formulation. According to our survey, some farmers still knew nothing about the payment policies of both the central government (28%) and local governments (37%) even after their implementation 9 years ago (Figure 5). The lack of policy knowledge and lack of direct participatory channels for policy-making caused a discrepancy between the current policy framework and the expectation of local farmers. This is exemplified, among others, in the expectations of farmers on payments. According to our survey in Liaoning Province, 96% of the local famers expected (much) higher payments for providing ecological services for their PBFs than the current 10 Yuan per mu (Figure 6). This can endanger the future success and legitimacy of the PBFs.

Referring to the responsibility for payments, around half of the respondents considered the central government solely responsible; 9% believed that the central and local governments are both responsible for the payment; and 30% considered that payments should be extracted from both the central government and the beneficiaries along the lower reaches of the watershed (residents and factories). Direct payment relationships between eco-service providers and beneficiaries are missing in current forest payment policies in China. Hence, local farmers understood the protection of PBFs, which mainly means stopping timber harvesting, as a duty to the central government, rather than to the local government or villages (not unlike what was found in other provinces; Chia, 2010; Liang, 2012b). The central government should in return satisfy their basic needs for livelihood. If the central government fails to provide sufficient support for their livelihood, it would be 'legitimate' for them to start timber harvesting again.

7.3 *Participation in the Demarcation of PBFs*

The participation of farmers in the demarcation of PBFs was to some extent greater than that in the policy-making process on PBFs and the payment schemes. In 2001, the PFD sent out professionals to help county forest bureaus to develop the general layout of PBFs according to the ecological importance of forestlands.

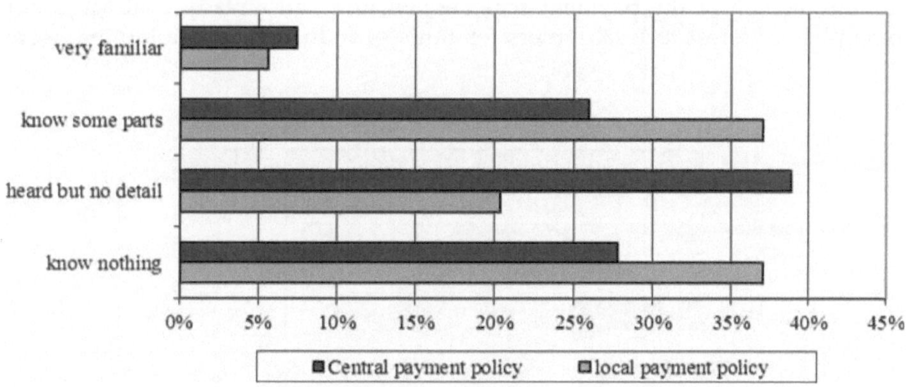

Figure 5. Farmers' knowledge on payment policies (n = 54).

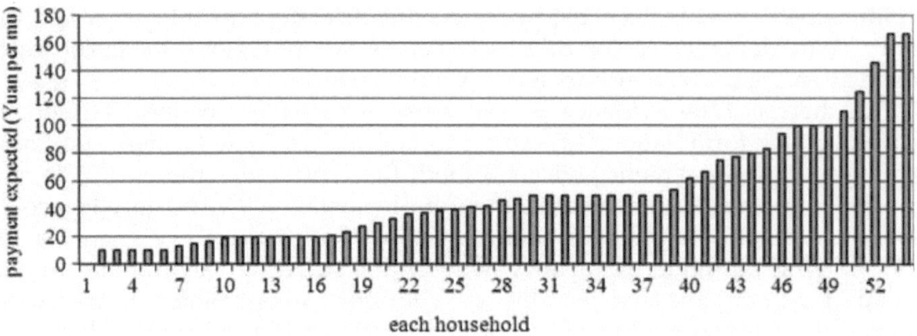

Figure 6. Farmers' expectations of payments for PBFs (Yuan per mu) ($n = 54$).

Updated satellite images and the data of forest resource inventories were employed to fix the boundaries of PBFs at each forestland compartment. After demarcation, forestry bureaus informed the villages about forestland use change and made management contracts with village committees. The demarcation was largely dominated by the PFD and county forestry bureaus and decided upon following scientific standards from the central and provincial forestry authorities.[14] Although consultation of village members was carried out by village committees through either all-member meetings or representative meetings, local communities and farmers did not have any final decisive power on which plot of forestland would be included into the PBFs. The 2001/2002 demarcation process was relatively smooth in Liaoning Province, benefiting from the collective tenure of forestland. Rather than negotiating with individual farmer households, the government only needed to reach an agreement with the village committees. According to the survey, 61% of farmer households were consulted during the demarcation of PBFs. The forms of consultation included all-member meetings (26%), representative meetings (37%), bulletins for public opinion (4%), and private negotiations (2%).[15]

The farmers showed different degrees of acceptance of the demarcation result: 26% of the farmers welcomed the decision, 28% of them basically accepted it, and 46% of them opposed the change of forestland into a PBF. Among the opponents, the main reasons for disagreement related to the reduced supply of fuel wood and the failure in fulfilling contracts with villages (Figure 7). Before the introduction of the payment schemes, villages had already contracted out some plots of forest to local famers for timber production. These farmers were

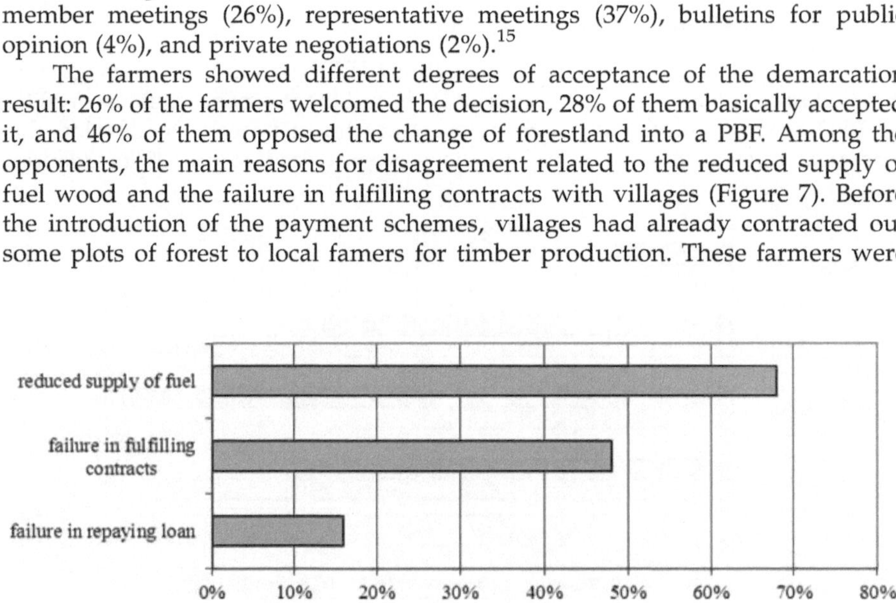

Figure 7. Reasons for opposing the demarcation of PBFs ($n = 25$).

obliged to pay rent to the villages for the use of collective forestland and had to repay the loan if they borrowed money from banks for forestry investment. Usually, these farmers strongly opposed the change of forestland into a PBF.

However, self-interest was not the only factor determining farmers' attitude towards the demarcation result. When discussing the ecological importance of PBFs, many farmers (80%) showed understanding and support for the demarcation of PBFs. This also demonstrates that to some extent local farmers can differentiate public good considerations from their private income concerns. In addition, a similar number of respondents thought that the process of demarcation was fair, in which each village and farmer household has been treated in the same way. Farmer households had different preferences for who should decide on the demarcation: governments (19%), village committees (28%), through negotiation with individual households (5%), and all-member village meeting (37%). A small group expressed no preference.

7.4 *Participation in the Management of PBFs*

Farmers also participated to some extent in the decisions on the management of PBFs. The management of PBFs in Benxi County and Xinbin County is carried out at three administrative levels: the county, township, and village levels. The county PBF management office is in charge of planning, coordinating, monitoring, and examining the management of PBF, of training foresters, and of distributing, managing, and monitoring the payment. The township PBF management stations are the main bodies for the management and protection of PBFs. The village committees help to monitor PBF management and evaluate the performance of forest rangers. After the county forestry bureau made an administrative responsibility contract with township governments, each village recommended local farmers for the position of forest rangers. Usually, this is a competitive process since the position provides a stable job with a relative high wage, compared with engaging in agricultural production (see Table 5). The township PBF management stations examine the qualification of the recommended candidates and decide on who will be employed as a forest ranger. The main task of forest rangers includes daily ranging in PBFs to prevent and control fire and pests; stopping illegal harvesting, occupation of forestland, digging and burrowing, hunting, and clearing of forest for cropping; reporting such illegal activities to local authorities; and other responsibilities detailed in the forest management contracts. The survey showed that 69% of the farmers were consulted before deciding on employing forest rangers. A large majority of the farmers (85%) indicated that they completely (41%) or basically (44%) agreed with the choice of forest rangers made by the village committee.

7.5 *Participation in the Examination of Payment Schemes*

According to the *Detailed Rules* (Liaoning PFD, 2006a), the examination of the performance of the payment schemes is carried out annually by municipal and county forestry bureaus together with financial bureaus. Initially, county forestry bureaus examine the performance of each township and report the result to municipal forestry bureaus. The municipal forestry bureaus verify the performance county by county and report to the PFD. Finally, the PFD organizes a recheck on a few selected counties. The examination includes the implementation of PBF management (compliance with regulations and laws, forest resources change, institution- and organization-building, information management, etc.), the use of the payment, and the implementation of PBF management contracts (ranging records, fire and pest control, preventing illegal harvesting and clearing the forest, etc.). Although the *Detailed Rules* do not require or mention the participation of local farmers in the examination, counties formulated and practised monitoring and management measures (e.g. *Benxi Manchu Autonomous County's Complementary Measures for Monitoring and Managing PBF Rangers*), which include participation of the local farmers. For instance, both county forestry bureaus set up permanent mailboxes in township forestry stations to collect public opinions on, among others, the performance of forest rangers in the management of the PBF. Moreover, the country PBF Management Office selected per village 20 (for small villages) or 30 (for large villages) local farmers for consultation about the performance of forest rangers. If a farmer ventilated gross misconduct of a forest ranger or less than 60% of the selected farmers were satisfied with the ranger's performance in PBF management, the ranger was dismissed from his post. According to the survey, 33% of the farmers had been consulted when forestry stations and higher level governments conducted the (re)examination of PBF management. In addition, 24% of the respondents have actively given opinions and revealed problems about the payment scheme implementation to village committees and higher level governments. Among them, 8% directed their input to county forestry bureaus, 69% to township forestry stations, and 54% to village committees.[16]

The survey showed that farmers had a clear preference for PBF performance examination by township forestry stations in combination with farmers (Figure 8). The existing mechanism of monitoring and examination of PES schemes basically corresponds with the preferences of farmers. Township forestry stations and county forestry bureaus play a pivotal role in the examination, with a substantial involvement of local farmers.

8. Conclusion

With the start of the third Millennium, China began to widely apply payment schemes for the protection and conservation of its forests. Huge financial resources have been allocated for the provision of ecological services throughout China. The implementation of the FEBCFP in Liaoning, the case study of this article, is far from an exceptional case. But China designed and implemented

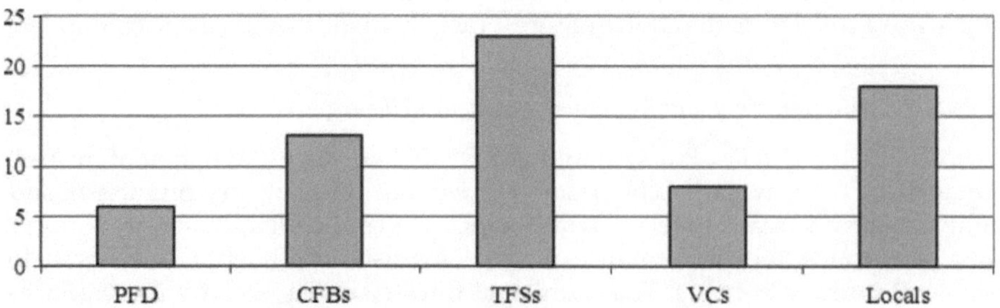

Figure 8. Preference for PBF examination body (in absolute numbers) ($n = 54$).
Note: PFD, provincial forestry department; CFBs, county forestry bureaus; TFSs, township forestry stations; VCs, village committees; and Locals, local farmers.

these innovative PES schemes for forest governance different from that of other countries.

Unlike a neoliberal paradigm where PES is thought to rely on the power of the market, China's payment schemes for forest ecological services mainly depend on the state in terms of scientific planning and zoning; monitoring, control, and enforcement; payments and incentives; and political will. Farmers regard the designation of PBF and the related payment schemes as a direct exchange with the state, in which in essence they simply are required to stop timber harvesting and the state guarantees basic livelihood requirements. This is quite different from the market-based PES in which a voluntary exchange takes place between the generation of ecological value and financial compensation. Farmers indicated that financial compensation was quite limited and a significant part of farmer support for PBFs was related to the possibilities of still carrying out commercial activities in these forests (such as agroforestry and harvesting of medicinal plants). In contrast to many other PES schemes, direct beneficiaries, such as factories and farmers and residents along the lower reaches of the watershed, have not been included in the payment schemes in Liaoning until now. More widely in China, forest payment schemes have been designed as single-sector narrow policies, without integration with other sectors. Moreover, the schemes have initially been set up centrally and implemented in a rather top-down mode. For quite some time, the central government has been the main coordinating, funding, and implementing agency of forest payment schemes. As such, these payment schemes for ecological services in PBFs seem to hardly resemble European models of political modernization in forest governance or the ideal–typical PES schemes as we know from, for instance, Costa Rica.

But that is only half of the story and analysis. In Liaoning, provincial and local forest payment schemes have emerged besides national ones, and through a process of decentralization, provincial and local governmental actors have become central players with significant 'room for manoeuvre' in fine-tuning and implementing PES. The reform of forest tenure from villages to farmer households also enhanced the involvement of non-state actors in these schemes, with their specific interest. Since the mid-2000s, we can identify a significant degree of participation of local farmers in the implementation (including the demarcation, management, and examination, but not so much in the policy-making) of the payment schemes and of the forestland tenure reform. By including PBFs in the collective forest tenure reform, farmers were also given a legitimate position to become active participants in the conservation and protection of forests. The decentralization and participation in PBF management, together with some room for farmers to carry out economic activities in PBFs, have resulted in a surprising support for PBFs, especially in the light of the rather low payments that farmers receive for environmental services. These tendencies of decentralization and participation resemble aspects of the idea of political modernization as articulated in European environmental governance, be it operationalized in a specific, Chinese, way. But channels for the participation of local farmers are still in need of expansion, especially in the demarcation of PBFs and in the more general policies on payments for environmental services.

Finally, it should be concluded that this 'Chinese mode of political modernization' seems to work. Provincial data and data from our case-study villages suggest that forest ecological improvements have been achieved, especially in the areas that are more vulnerable to soil erosion. And although further investigations are needed, it is quite likely that these improvements are related to the payment schemes. Moreover, although in Liaoning Province the PES schemes did affect the livelihood practices of rural farmers, changed the economic structure in these villages, and farmers considered the financial compensation to be too low, the schemes did not result in large-scale and major dissatisfaction among the farmers and problems with legitimacy for local governments implementing the schemes.

Notes

1. These programmes are the Natural Forest Protection Programme (NFPP), the Conversion of Cropland into Forest and Grassland Programme, the FEBCFP, the Sandification Control Programme in the Vicinity of Beijing and Tianjin, Shelterbelt Construction Programme, and the Wildlife Conservation and Nature Reserve Development Programme.
2. The reform also allows that a village as a whole keeps the use rights of its forestland as long as more than two-thirds of the village members agree with it at a formal member meeting. According to an interview with a local official of Liaoning Province, most of the villages preferred to redistribute the use rights of their forestland. Only in some villages, where collective forestry farms had a long successful history in forest management and benefit distribution, did farmers agree to maintain collective forest tenure with village committees.
3. Usually, if forest plots are too small and fragmented, farmers will delegate the task of forest protection to foresters.
4. In 2001, the central government enlisted Liaoning as a pilot province for the central fiscal payment scheme for forest ecological services with an annual funding of 105 million Yuan for 1.4 million hectares of forest.
5. In 2001, the province launched a project to ban timber harvesting for commercial purposes in natural forests in nine prime forested counties. This project, which adopted measures the same as those of the national Natural Forest Protection Project (NFPP), covered 787 thousand hectares of natural forest with 41.3 million Yuan for forest management and protection and 56.86 million Yuan transferred from the provincial finances to the county finances for compensation every year.
6. In 2004, following the formation of the national payment scheme, the province established its own provincial payment scheme and earmarked 18.27 million Yuan for protecting 406 thousand hectares of forest.
7. Different contract forms are possible: with individual households, with groups of households, with village clusters, with outside contractors, and with collectives. In Liaoning, contracts with individual farmers are by far the most dominant.
8. The full names of these two counties are Benxi Manchu Autonomous County and Xinbin Manchu Autonomous County.
9. Interview schedules and questionnaires are available in Liang (2012a).
10. Public benefit management at township level is implemented by teams of forest rangers in Liaoning Province. In principle, one forest ranger takes charge of managing 3000 mu of forests and one team leader is in charge of monitoring and checking the work of 10 forest rangers.
11. Some of them are the forest rangers and some are local farmers who succeeded in developing agroforestry, such as cultivating mushroom and Chinese herb medicine in forests.
12. The tenure reform clarified and substantiated the ownership of PBFs. It is expected to guarantee that individual farmer households as PBF owners can obtain the payment.
13. PPC is a forum for mediating policy differences between different parts of the government and the regions, and PPPCC is a political advisory body including broader members such as persons without party affiliation and mass organizations. Both are important institutions for public policy-making and consultation.
14. During interviews, local officials admitted that too much forestland had been demarcated into PBFs. The Liaoning PFD's standards are much more inclusive than the national standards. This resulted in a continuing controversy between the central government and the Liaoning provincial government. In 2009, the Liaoning PFD (2009) still insisted that the central government should accept its demarcation result (34 million mu), although only 27 million mu of PBF was confirmed

by the central government and provided with payment. This controversy reflects the self-interest of local governments in requesting central funding.

15. Some farmers have been informed and consulted through multiple forms.
16. Some farmers directed their opinions to village committees, forestry bureaus, and stations together.

References

Arts, B. & Buizer, M. (2009) Forests, discourses, institutions: A discursive-institutional analysis of global forest governance, *Forest Policy and Economics*, 11(5–6), pp. 340–347.

Bennett, M. T. (2009) *Markets for Ecosystem Services in China: An Exploration of China's 'Eco-compensation' and Other Market-based Environmental Policies* (Washington, DC: Forest Trends).

Blowers, A. (2002) Political modernisation and the environmental question: The case for strategic planning, *Local Government Studies*, 28(2), pp. 69–87.

Carter, N. T. & Mol, A. P. J. (2006) China and the environment: Domestic and transnational dynamics of a future hegemon, *Environmental Politics*, 15(2), pp. 331–345.

Chen, G. (2009) *Politics of China's Environmental Protection. Problems and Progress* (Singapore: World Scientific Publishers).

Chia, S.-T. E. (2010) Forest governance in Chinese villages: Community, the 'common interest', and common pool resources, PhD dissertation, Australian National University, School of International, Political and Strategic Studies, Canberra.

China Centre for Modernisation Research (2007) *China Modernization Report 2007: Study of Ecological Modernization* (Beijing: Beijing University Press).

Dai, L., Zhao, F., Shao, G., Zhou, L. & Tang, L. (2009) China's classification-based forest management: Procedures, problems, and prospects, *Environmental Management*, 43(6), pp. 1162–1173.

Guttman, D. & Song, Y. (2007) Making central–local relations work: Comparing America and China environmental governance systems, *Frontiers of Environmental Science & Engineering in China*, 1(4), pp. 418–433.

Jänicke, M. (1986) *Staatsversagen. Die Ohnmacht der Politik in die Industriegesellschaft* (Münich: Piper) [translated as State Failure. The Impotence of Politics in Industrial Society, 1990 (University Park: Pennsylvania State University Press)].

Jänicke, M. (1993) Über ökologische und politische Modernisierungen, *Zeitschrift für Umweltpolitik und Umweltrecht*, 16(2), pp. 159–175.

Jiang, H. (2006) Decentralization, ecological construction, and the environment in post-reform China: Case study from Uxin banner, Inner Mongolia, *World Development*, 34(11), pp. 1907–1921.

Kotilainen, J., Tysiachniouk, M., Kuliasova, A., Kuliasov, I. & Pchelkina, S. (2008) The potential for ecological modernisation in Russia: Scenarios from the forest industry, *Environmental Politics*, 17(1), pp. 58–77.

Liang, D. (2012a) *Payment Schemes for Forest Ecosystem Services in China: Policy, Practices and Performance* (Wageningen: Wageningen University Press).

Liang, D. (2012b) Evaluating payment schemes for forest ecological services in China: The case of Fujian Province, *Forest Policies and Economics*.

Liaoning Provincial Forestry Department (2006a) *Liaoning Province's Detailed Rules for Implementing the Management of Public Ecological Benefit Forest* (Shenyang: PFD).

Liaoning Provincial Forestry Department (2006b) *Liaoning Provincial Forestry Department's Report on Prior PBF Management and Examination in 2006* (Shenyang: PFD).

Liaoning Provincial Forestry Department (2009) *Opinion on Facilitating Reform on the Management and Protection Mechanism on PBF* (Shenyang: PFD).

Liaoning Provincial Government (2005) *Liaoning Provincial Government's Notice on Deepening Collective Forest Tenure Reform* (Shenyang: Liaoning Provincial Government).

Miranda, M., Porras, I. T. & Moreno, M. L. (2003) *The Social Impacts of Payments for Environmental Services in Costa Rica: A Quantitative Field Survey and Analysis of the Virilla Watershed* (London: International Institute for Environment and Development).

Mol, A. P. J. (2002) Political modernisation and environmental governance: Between delinking and linking, *Europæa. Journal of the Europeanists*, 8(1–2), pp. 169–186.

Mol, A. P. J. (2003) Joint environmental policy-making in Europe: Between deregulation and political modernization, *Society and Natural Resources*, 16(4), pp. 335–348.

Mol, A. P. J. (2006) Environment and modernity in transitional China: Frontiers of ecological modernisation, *Development and Change*, 37(1), pp. 29–56.

Mol, A. P. J., Sonnenfeld, D. A. & Spaargaren, G. (Eds) (2009) *The Ecological Modernization Reader: Environmental Reform in Theory and Practice* (London: Routledge).

Rosenbaum, W. A. (2000) Escaping the 'battered agency syndrome': EPA's gamble with regulatory reinvention, in: N.J. Vig & M.E. Kraft (Eds), *Environmental Policy*, 4th ed., pp. 165–189 (Washington, DC: CQ Press).

Sánchez-Azofeifa, G. A., Pfaff, A., Robalino, J. A. & Boomhower, J. P. (2007) Costa Rica's payment for environmental services program: Intention, implementation, and impact, *Conservation Biology*, 21(5), pp. 1165–1173.

State Forestry Administration (2006) *Investigation Report on Collective Forest Tenure Reform. Department of Policy and Regulation* (Beijing: SFA).

van Tatenhove, J. & Leroy, P. (2003) Environment and participation in a context of political modernisation, *Environmental Values*, 12(2), pp. 155–174.

van Tatenhove, J., Arts, B. & Leroy, P. (Eds) (2000) *Political Modernisation and the Environment. The Renewal of Environmental Policy Arrangements* (Dordrecht: Kluwer).

Veenman, S., Liefferink, D. & Arts, B. (2009) A short history of Dutch forest policy: The 'de-institutionalisation' of a policy arrangement, *Forest Policy and Economics*, 11(3), pp. 202–208.

Wang, X. (2008) Analysis on cause and characteristics of soil erosion in Liaoning Province, *China Science and Technology Information*, 12, pp. 24–25.

Wunder, S., Engel, S. & Pagiola, S. (2008) Taking stock: A comparative analysis of payments for environmental services programs in developed and developing countries, *Ecological Economics*, 65(4), pp. 834–852.

Xu, J., White, A. & Lele, U. (2008) *China's Forest Tenure Reforms. Impacts and Implications for Choice, Conservation, and Climate Change* (Washington, DC: Rights and Resources).

Zbinden, S. & Lee, D. R. (2005) Paying for environmental services: An analysis of participation in Costa Rica's PSA program, *World Development*, 33(2), pp. 255–272.

Zhang, K. & Wen, Z. (2008) Review and challenges of policies of environmental protection and sustainable development in China, *Journal of Environmental Management*, 88(4), pp. 1249–1261.

Zhang, H., Zhimu, D., Zhang, X., Zhang, J., Wang, W. & Cao, Y. (2005) Comparative analysis of management techniques and administrative policies of public welfare forests in Japan and Liaoning, *Journal of Liaoning Forestry Science & Technology*, 5, pp. 35–40.

Zhang, L., Mol, A. P. J. & Sonnenfeld, D. A. (2007) The interpretation of ecological modernization in China, *Environmental Politics*, 16(4), pp. 659–668.

Zhang, L., Tu, Q. & Mol, A. P. J. (2008) Payment for environmental services: The sloping land conversion program in Ningxia autonomous region of China, *China & World Economy*, 16(2), pp. 66–81.

Zheng, H. & Zhang, L. (2006) *Chinese Practices of Ecological Compensation and Payments for Ecological and Environmental Services and its Policies in River Basins* (Washington, DC: World Bank).

Zhong, L. & Mol, A. P. J. (2008) Participatory environmental governance in China: Public hearings on urban water tariff setting, *Journal of Environmental Management*, 88(4), pp. 899–913.

Environmental Activism in Provincial China

FENGSHI WU

Department of Government and Public Administration, Chinese University of Hong Kong, Shatin, N.T, Hong Kong SAR

ABSTRACT *Environmental activism and non-governmental organizations (NGOs) are not limited to major cities, but have gradually taken root in provincial and local politics in China. This paper compares the development and characteristics of environmental activism in Guangdong province and Guangxi autonomous region, with the aim of shedding light on the causes of regional disparities in this field. With extensive and strong ties to peer organizations across issue areas within the province, grassroots environmental groups in Guangdong have grown as an integral part of local civil society. They have survived, through mutual-support, on a larger scale than those in Guangxi. Also, with a higher level of networking and capacity, environmental NGOs in Guangdong are more able to utilize the new opportunities offered by recent institutional reforms in social management and environmental protection in the province. The paper, thus, challenges the argument that over-emphasizes the domination of the state and points out that inter-organizational partnerships and the overall maturity of the civil society sector contribute most to the uneven development of grassroots environmental activism.*

Introduction

The past 20 years have seen public environmental awareness grow in China, and citizens are more prone to take action against industrial pollution. Not only are protests held by pollution victims across the country, but concerned citizens have also started to attend public consultation sessions related to major construction projects, and established non-governmental organizations (NGOs) to voice their disagreements with governmental decisions. The environmental field arguably involves the most visible activist community, largest number of grassroots NGOs, and has the longest history of public policy advocacy across issue areas in the country (Hildebrandt, 2011). An increasing body of literature has emerged on the rise of environmental activism driven by NGOs in China. With few exceptions, literature on the topic to date has relied primarily on detailed case studies of individual NGOs and campaigns, yet arrived at generalized conclusions at the national level. The concepts of a 'green civil society' and an 'environmental movement' have been envisioned and argued in contrast to the Chinese state/regime as a whole (Sun & Zhao, 2008; Turner, 2004). However, the field lacks systematic knowledge of the regional spread of, and disparities within, such a movement.

This paper examines the environmental NGO-driven activism at the provincial level and highlights inter-organizational connections as an important characteristic of the sector. It defines NGO-driven activism in contrast with grievance-driven social mobilization. While the latter is usually a result of individual or community victimhood and sets out primarily to win compensation, NGOs tend to organize their activism with long-term goals, promoting the public good, and are aware of policy implications. Most environmental NGOs in China see their role as that of engaging in self-limited, rational bargaining with various state agencies in the long-run (Buesgen, 2008; Cooper, 2006; Hildebrandt, 2011; Spires, 2011; Tang, 2008; Turner, 2004; Yang, 2005; Yang & Gabriel, 2001).[1]

In theory, NGOs are voluntary-based, not-for-profit, private and civic organizations that exhibit a minimum level of self-governance (Salamon & Anheier, 1997, pp. 33–34). This paper is cautious when applying the above definition to study environmental NGOs and adds two specific modifications: First, due to the fact that the current laws in China make it extremely difficult to obtain the official status of non-profit organization, this paper relaxes the issue of formal status and focuses on the actual work and organizational nature when determining whether a grassroots group is an NGO or not. It considers any legally registered or publicly recognized social organization to be an NGO, as long as it, in practice, fulfils the requirement of being voluntary-based, not-for-profit and autonomous. Second, it focuses on groups established outside of the state apparatus and differentiates them from quasi-governmental, state-affiliated organizations or government-organized NGOs (Deng, 2010; Hildebrandt, 2011; Spires, 2011; Wu & Chan, 2012).

This paper consists of four main parts. The first introduces the current state of development of environmental activism and NGOs in China and reviews the existing literature on how to understand such a phenomenon and possibly explain the variations at the provincial level. The second part presents the main characteristics of the environmental NGO communities in Guangdong province and Guangxi autonomous region. The remaining two parts examine the most relevant factors that can explain the differences between the two cases. They find that environmental NGOs have drawn substantial strength from the support of kindred organizations. Horizontal synergies between civic groups stand out as an indispensable factor for a vibrant environmental activism community in Guangdong. With support from peer organizations, environmental NGOs in Guangdong seem to have benefited from the recent governmental initiatives of enhancing public participation in environmental governance.

The choice of comparing the two southern provinces in China is deliberate, in order to search for meso-level explanations for regional variation in environmental NGO development. An examination of these two cases can offer insights on which crucial factors have helped environmental NGOs in some places, but not others, survive and thrive. This research conducted extensive field work and interviews in the two provinces during the periods of May–July 2008,

November 2010, February 2011 and March 2012. In addition to semi-structured interviews, this research also employed three focus-group discussions with NGO staff, activists and journalists (Appendix). In Guangdong, the first focus-group discussion was conducted with two investigative journalists and two public intellectuals and professors (GD Group 1); and, the second group involved one staff member each from an environmental NGO and a platform NGO (GD Group 2). In Guangxi, two current and one former local environmental NGO staff members participated in the focus-group discussion (GX Group).

Regional Patterns of Environmental NGO Development in China

The rise of modern environmental activism in China has gone through three main phases (Wu, 2009). It first emerged in China in the 1980s, driven by a few leading activists, scientists and journalists, and was carried out within the social elite and high-level policy-making circles. The second phase was marked by the emergence of environmental NGOs in major cities established by prominent activists, and several successful NGO-initiated public environmental campaigns (Economy, 2004; Yang, 2005). The NGO community's collective response to the Sichuan earthquake in May 2008 triggered the emergence of a third phase of environmentalism in China: a younger, less- recognizable, yet more diverse generation of social entrepreneurs and NGO leaders are taking the centre stage of environmental activism (Shieh & Deng, 2011). 'Environmental NGO' has now become a nationwide phenomenon, and has spread to all provinces in the country.

Based on three comprehensive surveys, this paper has mapped the development of environmental NGOs by province in China.[2] Although Figure 1 may not

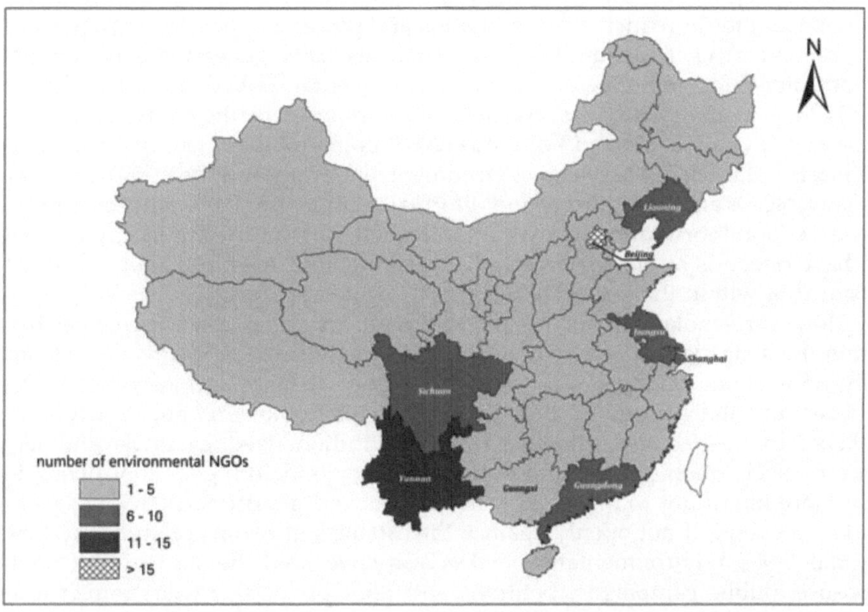

Figure 1. Number of grassroots environmental NGOs in Mainland China by province.
Sources: Directory of Environmental NGOs (2005), Map of Chinese Environmental NGOs (2009), and NGOCN Online Database of Environmental NGOs (2010).

tell the whole story since individual environmentalists and cross-regional networks are difficult to illustrate, what it demonstrates is the provincial distribution of the most active grassroots green NGOs. The provinces of China can be divided into three tiers, measured by their number of grassroots environmental NGOs. In Beijing municipality and Yunnan province, the environmental NGO sectors have grown biggest and arguably reached highest level of maturity in the country, which is supported by existing cases studies (Cooper, 2006; Hayes, 2007; Turner & Wu, 2001; Wu, 2009). All other provinces are divided into two tiers: in the four provinces of Liaoning, Sichuan, Jiangsu, and Guangdong and the municipality of Shanghai, the surveys find a relatively more visible environmental NGO presence, with 6–10 groups; while, in the remaining 23 provinces, environmental NGOs exist on a very limited scale, where no more than five organizations can be identified.

Within the existing literature, there are various theories on the causes of the rise of environmental activism in China, such as the theses of ecological modernization and the greening of the state at the macro-level (Carter & Mol, 2007; Ho & Vermeer, 2006; Mol, 2006; Mol & Carter, 2006) and the micro-level factors of *guanxi* and organizational leadership (Xie, 2009). External factors, particularly the financial support and expertise provided by international environmental NGOs, best explain the cases of Beijing and Yunnan as champions of local environmental activism until the early 2000s (Chen, 2010; Morton, 2005; Xie, 2011). Yet, the challenge remains of sorting out the meso-level factors that are specific to the Chinese *politique* and have directly contributed to the provincial-level disparities shown in Figure 1.

Governing structures and institutional arrangements matter to social activism in general in China. Based on a comparative study of three movements opposing large dam projects in the late 1990s and the early 2000s, Mertha (2008) argues that fragmented inter-bureaucratic coordination and power competition between large construction firms and administrative agencies have generated new political opportunities for the emergence of environmental policy advocacy led by NGOs, investigative journalists and activists. Supported by the many contributors of the edited volume, Ho and Edmonds (2008) point out that there is a 'negotiated symbiosis' relationship between environmental activists and the authorities, and, in a way, NGOs and their formal and informal relations are inevitably 'embedded' in the 'semi-authoritarian context'. In a nutshell, how much autonomy an NGO can have depends on how successfully it can mobilize resources, not necessarily outside, but within the extensive webs of the state apparatus.

However, scholarship such as the above—focusing on how the changes from within the state system can enable and shape activism—implicitly, yet substantially, underestimates the self-driving character of activists and advocacy NGOs. Activists and their teams by definition are 'whistle blowers' and 'movers and shakers'. In the Chinese context, where institutionalized channels and legal space for social mobilization and policy advocacy remain highly controlled, it is even more important to recognize that activists and grassroots NGO leaders are working beyond, if not openly against, the established realms of authority. Since the mid-1990s, environmentalists and NGOs have tried the strategies of media exposure, public campaigns, petitions and pilot projects to exert influence on policy-making and implementation, at a time when no specific law was enacted or official speeches by a top leader were given endorsing the ideas related to policy advocacy and public participation. The policy concept of 'social development (*shehui jianshe*)' was only officially endorsed by the Chinese Communist Party the first time in Hu Jintao's plenary speech at the 17th National Congress in 2007.[3] According to several keen observers, leading environmentalists and NGOs were pioneers in voicing different policy opinions and channelling social demands from the bottom up, and they took high political risks in 'pushing the envelope' and testing the 'red lines' of activism and advocacy (Economy, 2004; Saich, 2000; Turner, 2004; Turner & Wu, 2001). It is, thus, one-sided to argue that the activists and NGOs thrive on the sympathy and consent of the state, rather than by their own innovations, the legitimacy they have earned from the communities they serve, and other opportunities they obtained independently.

The search for the root cause of the emergence and continual growth of Chinese environmental NGOs has to consider the power and capacity of civil society generated from within. Comparative studies on social coalitions and political change have pinpointed the importance of inter-organizational coordination and trust-building within the civil society sector to resisting state penetration (Carroll & Ratner, 1996; Obach, 2004; Rochon & Meyer, 1997; Van Dyke & Mccammon, 2010; Weiss, 2006). In mature democracies, it is found that collaborative ties among NGOs working in the public interest form the basis of 'horizontal solidarity' and integrate the civic power across existing social divides (Baldassarri & Diani, 2007). Inter-organizational connections become more crucial for resource mobilization, risk-sharing and collective action under non-democratic regimes. The experiences of the former Soviet Bloc in Central Europe explain that because there was no formal channel for dissident voices and direct confrontation could cause violent repression, civil society organizations relied on each other to engender an 'explicitly autonomous and implicitly alternative' political sphere. Constructing, or 're-inventing', such a condition of 'parallel politics' was critical to conserve the momentum of social activism and protect many grassroots initiatives from being crushed at the early stage of political transformation (Goldfarb, 2001; Kubik, 1994; Matynia, 2001; Tismaneanu, 1990, 1993).

Drawing upon the theorizations of the power and capacity generated within civil society, this paper conceives that grassroots environmental NGOs in China are essentially nurtured by their social surroundings, particularly the networks of peer organizations and activists, instead of their inevitable relations with the state. Most of the first generation of environmental NGOs shared an organizational genealogy and are related to the Beijing-based Friends of Nature (FoN) and Green Camp (Economy, 2004; Ru & Ortolano, 2009). It is not exaggeration to point out that they were born into a network of similar organizations. The leaders of this group of NGOs were inspired by the idea of 'NGO' and were encouraged to pursue it precisely because of peers, not the state. As environmental activism further developed in the late 1990s, such networks of environmental NGO extended beyond personal relations and included more partnerships in organizing public campaigns, bargaining with state agencies and research (Buesgen, 2008; Sullivan & Xie, 2009). The 2008 Sichuan earthquake and the immense need for relief and reconstruction efforts urged the environmental NGO community to act together and in turn led to more autonomy from the state (Shieh & Deng, 2011).[4]

Suggested by the above comparative theories and preliminary empirical evidence of environmental activism China, this research explores how the contingent socio-political settings have shaped the overall development of environmental

NGOs with a focus on the patterns of inter-organizational relations within the civil society sector. To do so, it further differentiates the vertical from horizontal relations among all NGOs. 'Vertical networks' include an NGO's connections with other peer organizations outside the province and indicate not only geographical distance, but also a hierarchy in capacity. Usually, the external NGOs play the role of main donor and financial supporter to the local ones. 'Horizontal networks', in contrast, include an NGO's connections with peer organizations within the province and imply more equal exchanges and mutual-support. Such exchange and support are usually not in the form of funding, but information and knowledge, social capital and moral encouragement. The analysis of both kinds of networks among NGOs in Guangdong and Guangxi will reveal different effects of them on the overall development of environmental activism at the provincial level. Beijing and Yunnan by and large are outliers in Figure 1, and a comparative study of Guangdong and Guangxi better serves the purpose of explaining the uneven trends of NGO-driven environmental activism across the second and third tiers, which encompass the majority of Chinese provinces. Being of the tier-two provinces, Guangdong particularly offers insights to understand how environmental NGOs in some provinces, not others, can overcome the structural barriers and political surveillance at large, and gradually build up a coalition of peer organizations to push forward the environmental agenda.

Environmental Activism and NGOs in Guangdong and Guangxi

In the overall picture of environmental activism in China, both Guangdong and Guangxi are latecomers. Guangzhou, the capital city of Guangdong province, is a major economic centre of national importance, but green NGOs did not emerge there until the early 2000s. Similar in terms of its rich ethno-cultural diversity and poor economic performance, Guangxi autonomous region cannot yet compete with neighbouring Yunnan province, in terms of the growth and proliferation of green NGOs. Unlike Beijing, Yunnan or Sichuan (after the earthquake in 2008), neither Guangdong nor Guangxi can identify 'environmental heroes' who could have symbolic public influence, nor a leading green NGO comparable with FoN in Beijing or the Grassroots Community (*Re'ai Jiayuan*) in Shanghai, which has served as a platform for the green community to come together and develop bonds. Some of the earliest green NGOs in the two provinces were initiated under the direct influence and support of leading NGOs and veteran activists in Beijing or other major cities. However, after the initial take-off, Guangdong has seen a more steady growth of environmental NGOs, and they together present a more diversified agenda for public education, campaigning and policy advocacy.

Five nationwide environmental award lists were used to search for key environmental activists and early histories of NGO development in the two provinces.[5] Only four award winners were found from each province (Table 1). Professor Pan Wenshi of Beijing University is arguably the only environmental 'celebrity' in the two provinces. Known as the world's leading field biologist and expert on the giant panda, Pan established a biodiversity research and conservation centre for the white-headed langur in Chongzuo prefecture, southwest Guangxi, in 1996. Based on this centre, he registered the Chongzuo Ecological Park as an NGO at the local authority in 2006. Nevertheless, Pan has been better known to the world outside Guangxi than locally, and his NGO has not

Table 1. Search for key environmentalists and environmentally active NGOs in Guangdong and Guangxi

	Guangdong	Guangxi
2009	(1) Green Pearl River (SEE TNC Award)	
2008		(1) Chongzuo Ecological Park (SEE TNC Award)
2007		(2) Xu Xinbang, Fangchen District, Fangchengang City (China Environmental Protection Award)
2006		(3) Chongzuo Ecological Park (Ford Award)
2003	(2) Su Yongshan, teacher of the Taoyuan School of Foshan (Earth Award) (3) Xia Tao, school teacher, affiliated with the South China Normal University (Earth Award)	
2002		(4) GMCGS (Earth Award)
2001	(4) Ke Wenzhong, Chief of Guangzhou Office, China Environmental News (Earth Award)	(5) GMCGS (Ford Motor Award)
2000		(6) Li Li shows host of Guangxi TV station (Earth Award)

yet played a leading role in environmental activism in the province. Established in the late 1990s, the Guangxi Medical College Green Salon (GMCGS), winner of the Earth Award and the Ford Motor Award, was once among the most active college green groups in the whole country.[6] To some extent, these early award-winning initiatives by ecologists and college students in the 1990s could have developed further and led to the blossoming of environmental activism in Guangxi, yet the momentum was missed and it took almost another decade for more green NGOs to emerge in the province.

Environmental NGOs gradually emerged and increased in both Guangdong and Guangxi in the early 2000s. The existing surveys have identified at least seven established NGOs in Guangdong and two in Guangxi. This research has found two more NGOs in each province: the Bird-Watching Society in Shenzhen and Biker Guangzhou in Guangdong, and the Dragon Project Organic Farm and the Wild Flower Association in Guangxi. Table 2 lists all these NGOs, their years of establishment and main missions.

As shown above, the environmental NGO community in Guangdong as a whole emerged earlier than that in Guangxi, with seven NGOs established by 2005. Also, the environmental NGO community in Guangdong is spreading faster. Three main cities in Guangdong—Guangzhou, Shenzhen and Dongguan—have all seen the rise of public awareness and NGOs addressing specific environmental related issues in the city. Groups such as Biker Guangzhou, Green Point (GP), the 'Blue Ocean' Conservation Association and the Green Pearl River have built up networks outside their home cities and are concerned with problems in other parts of the province and beyond. Despite being one of the youngest NGOs in Guangdong and the country, Biker Guangzhou has grown rapidly and has become a hub of environmentalists and young activists—turning its own office into a meeting place for all environmental NGOs in Guangzhou city and concerned journalists and experts from all around the country.

Table 2. Environmental NGOs in Guangdong and Guangxi

Guangdong		Guangxi	
NGOs (year of establishment)	Working focus	NGOs (year of establishment)	Working focus
(1) Biker Guangzhou (2009)[a]	Green lifestyle, climate change	(1) Chongzuo Ecological Park (2006)	Conservation and community sustainable development
(2) Dongguan Green Volunteers (2005)	Local environmental protection education and voluntary activities	(2) Dragon Project Organic Farm in Yangshuo County (2008)[a]	Community sustainable development
(3) FoN Shenzhen Members' Group (2003)	General environmental issues in the province	(3) Guangxi Green Alliance (2007)	Environmental education and protection in general
(4) Green Pearl River in Dongguan (2003)	River conservation in the province and the Pearl River Delta	(4) Nanning Wild Flower Association (2011)[a]	Environmental education and public awareness
(5) GP Guangzhou (2007)	Environmental public education, general environmental concerns in the province		
(6) Green Rivers Shenzhen Members' Group (2005)	River and nature conservation in the province and the country		
(7) Shenzhen 'Blue Ocean' Conservation Association (2002)	Protection of mangrove and promotion of public environmental awareness		
(8) Shenzhen Bird-watching Society (2004)[a]	Promotion of environmental awareness		
(9) South China Nature Society (2005)	Endangered species, local conservation		

Sources: Directory of Environmental NGOs (2005), Map of Chinese Environmental NGOs (2009), and NGOCN Online Database of Environmental NGOs (2010).
[a]Based on field work conducted in 2011–2012.

Moreover, the scope of Guangdong environmental NGOs' work is growing broader, and some NGOs have reached a high level of specialization. Besides a green lifestyle, bird-watching and general public awareness, an increasing number of environmental NGOs in Guangdong have started working on new and compound issues such as climate change and coastal conservation. With accumulated field experiences, Green Pearl River and Blue Ocean have both grown into expert groups in their own areas. In particular, Blue Ocean, as the only NGO among the very few specialized institutions in the country focusing on the conservation of mangrove and continental shelf ecosystem, is gaining reputation internationally.

In contrast, not only are there fewer green NGOs in Guangxi, but also the geographic spread and professional scope of their work are much more limited. Like

Professor Pan and Chongzuo Ecological Park, the Dragon Project Organic Farm in Yangshuo county was started by a non-local—a retired business executive from Beijing. The Project team has been working together with three villages of one township in promoting organic farming techniques since 2008. Both Guangxi Green Alliance and Wild Flower Association are based in Nanning, the capital of Guangxi, and focus on raising public environmental awareness by educational programmes and organizing eco-study tours. Green Alliance works with a relatively broader environmental mission and intends to become a provincial-wide network, but its outreach has not gone beyond main universities and youth-based groups.

It also appears that the organizational missions of Guangdong environmental NGOs reflect the kinds of principles that are close to the 'post-materialism' conceptualized by Inglehart (1977) and examined by the World Values Surveys. During the year of preparations for the Guangzhou 2010 Asian Games, at least two public campaigns were staged in the province. One, against a local radio station's decision to change the programmes in the local dialect, Cantonese, to standard Chinese (*putonghua*) in order to make the cultural scenes of the city look more 'modern and open'; and, the other, calling for scaling down the city's nighttime light-display—specially designed for the Asian Games—in order to save electricity. Considered in the general context of China, these public campaigns are progressive in terms of the ideas they promote. Similarly, the missions of many NGOs in Guangdong studied in this research also reflect local citizens' non-materialist demands from the authority. While such post-materialist goals have a growing hold on public attention in Guangdong, Guangxi's environmentalists still regard helping rural villagers with income generation and associating pollution with resource loss as indispensable and the most effective strategy in winning popular support for their work.

On first inspection, student environmental groups and non-governmental initiatives in species conservation sprouted in both southern provinces in the late 1990s, but it was not until the arrival of the new millennium that establishing NGOs became a common strategy for environmentalists to pursue their goals, resulting in an increase in the number of environmental NGOs. Variation in socio-economic development sheds some light on the different conditions for the initial patterns of environmentalism NGOs. Continuous economic growth and urbanization, coupled with improvements in basic infrastructure, mass media and education since the end of the 1970s, have undoubtedly contributed to the change in public perceptions on the meaning of development and modernization in Guangdong (Table 3). Thus, it is not particularly surprising to see the

Table 3. Development of media and higher education in Guangdong and Guangxi

	Guangdong	Guangxi
GDP (billion RMB)	3067	589
Population (million)[a]	80	50
Urban population/total	63%	33%
Hospitals and medical clinics	16,328	10,654
Number of newspapers	100	55
Number of institutions of higher education	129	79
Number of students enrolled in higher education (10,000)	133	55

Source: Provincial statistics books of Guangdong and Guangxi (2010).
[a]Registered population (*huji renkou*).

earlier rise of Guangdong environmental NGOs, particularly in the main cities, and their organizational missions being different from their peers in Guangxi. However, it is not clear yet how the environmental NGO community in Guangdong has kept its momentum, continued to grow, covered more issues and become experts in their own fields. The next two sections present more differences exhibited by the two groups of environmental NGOs and explain the factors that have led to the diverging trajectories.

Horizontal and Vertical Synergies within Civil Society

To further identify factors that have directly contributed the development of environmental NGOs, this research examined NGO partnerships and peer relations with interview notes and published organizational information (Table 4). Two crude measurements are used to denote the relative level of intensity of inter-organizational connections: the total number of peer organizations that an environmental NGO is in regular contact and collaboration with and the percentage of annual budget that is from an international NGO. 'Peer organizations' include both international and domestic grassroots NGOs that work in environmental protection or related issues. 'Low' refers to between 1–5, 'Medium' 6–10 and 'High' above 10. The majority of environmental activists from both provinces concurred that networking is essential to their work and capacity-building; however, their organizations differ in overall patterns of peer relationship.

The first interpretation of Table 5 is that networking with peer organizations is crucial to any grassroots environmental NGOs in Guangdong and Guangxi. No local NGO is found surviving in complete isolation. Most of the dozen

Table 4. Environmental NGOs' vertical and horizontal networks

	Vertical networks (outside the province)		Horizontal networks (within the province)
Environmental NGOs	International NGOs	Chinese NGOs	Local NGOs
Guangdong			
Biker Guangzhou	Medium	Medium	High
Dongguan Green Volunteers	Low	Low	Medium
FoN-Shenzhen Members' Group	Low	Medium	Medium
Green Pearl River	Low	High	Medium
GP Guangzhou	Medium	High	High
Green Rivers Shenzhen Members' Group	Low	Medium	Medium
South China Nature Society	Low	Medium	Medium
Shenzhen Bird-Watching Society	Low	Medium	High
Shenzhen 'Blue Ocean' Conservation Association	Medium	High	Medium
Guangxi			
Chongzuo Ecological Park	High	High	Low
'Dragon Project' Organic Farm	Medium	Medium	Low
Green Alliance of Guangxi	Medium	Medium	Medium
Wild Flower Association Nanning	Low	Medium	Low

Note: Low = 1–5; medium = 6–10; high = more than 10.

Table 5. 2010 PITI rankings for all selected Guangdong and Guangxi cities

Guangdong	Ranking	Guangxi	Ranking
Shenzhen	14	Nanning	34
Guangzhou	25	Guilin	59
Foshan	25	Beihai	79
Zhongshan	28	Liuzhou	96
Shantou	30	Zhanjiang	111
Dongguan	42		
Zhuhai	45		

Note: 2010 PITI was published jointly by the Institute for Public and Environment in Beijing and the US National Resource Defense Council's Beijing Office. The PITI project selected and surveyed 113 cities across the country with regard to municipal EPBs' implementation of the regulation on pollution information disclosure.

NGOs have woven extensive inter-organizational networks with international and domestic NGOs, particularly with experienced environmental NGOs in Beijing, Sichuan and Shanghai. It is evident that environmental NGOs in the two southern provinces are part of the overall rise of environmental activism in China. Some of them are key links of cross-regional networks; for example, the Green Pearl River is part of the network focusing on water conservation in the whole country, and Blue Ocean on coastal conservation in southern China. Some were established as a direct result of the strong encouragement from leading environmental NGOs or veteran activists, such as the Shenzhen members groups of FoN (Beijing) and Green Rivers (Sichuan province).

The second, and more important finding from mapping environmental NGOs' networks, is the uneven development of vertical and horizontal peer networks across the two cases. While the environmental NGO sector in Guangdong is better integrated within the province, connections with NGOs from outside of the province are more prevalent and important in Guangxi. The two NGOs in Chongzuo and Yangshuo are still, in a sense, 'parachute initiatives'—they were started by activists from Beijing and have only gradually taken root in local communities. The Wild Flower Association was established by a local environmentalist who was heavily involved in student environmental activism and the GMCGS. Its funding at this point is all from international NGOs. Green Alliance of Guangxi seems to be different from its peers in the province and more akin to the green NGOs in Guangdong.[7] It was established by a group of young, urban college degree-holders in Nanning city in 2007. Most of the founding members were already active during their college years and involved in the GMCGS, which was closely connected with the green NGOs in Beijing at the time. However, due to funding shortages, all Green Alliance staff worked on a part-time basis and the organization remained mostly dormant until a large grant was received from an international NGO in 2009. There are frequent exchanges between the Wild Flower and Green Alliance, thanks to their ties with GMCGS. However, this is not the case for all environmental NGOs in Guangxi, and the two rural ones are almost disconnected from the rest of the province's activist community.

The ecology of Guangdong environmental NGOs is quite different from that in Guangxi. At least three environmental NGOs are making conscious efforts to integrate all grassroots environmental activism initiatives in Guangdong. GP, as the most experienced green NGO in South China, has always been supportive to new peer groups. Biker Guangzhou is quickly becoming a meeting point for young environmentalists and college green groups. And, with its specialty in mangrove protection, Blue Ocean in Shenzhen is growing into a stimulating organization, both attracting capable environmentalists and supplying expert advice to other conservation initiatives along the southern coastal line.

Unlike their peers in Guangxi, Guangdong environmental NGOs are better connected and integrated with other NGOs in the province than with those from outside. As Table 4 illustrates, with very few exceptions, they most frequently communicate, consult and partner with local groups. This is not to say that they are not open to international collaboration and assistance; rather, it is to point out that most of them do not depend on foreign donors or partners. Six of nine NGOs lack extensive experience with international NGOs. Dongguan Green Volunteers and South China Nature Society are two community-based groups initiated by local activists that primarily focus on local issues. Different from the Chongzuo Ecological Park and the Dragon Project, these two have mobilized most resource from within the province.

Even though Guangdong is a latecomer in terms of environmental activism, the province has, since the end of the 1970s, hosted quite a large number of grassroots social groups in the fields of independent union mobilization, migrant workers' rights, patients' mutual-support and rights of the disabled (Spires, 2007). The province has seen the revival of the sector of private charities that are highly active in social service delivery. Since 2005, one organization in particular, the Institute for Civil Society (ICS) at the Sun Yat-Sen University in Guangzhou city, has played a crucial role in bringing various grassroots voluntary groups together and generating extra thrust for civil society development in the province. ICS's founding members include Professors Chan Kin Man (from Hong Kong) and Zhu Jiangang who are now among the best-known public intellectuals advocating for the development of civil society in China.[8] Every year, ICS provides dozens of training opportunities for various grassroots social groups in Guangdong and beyond. Since its inception, ICS has gradually constructed a dense web of collaboration among all social activism groups in Guangdong and the environmental field is no exception. Although the green groups in the province may not know each other well, they have all been in contact with ICS. More importantly, most see themselves as part of a civil society, or *minjian liliang* (social forces), in Guangdong. They share with, and learn from, NGOs from other fields in terms of outreach strategies, public education, community mobilization and policy advocacy (GD Group 2). For example, on 12 January 2010, a month after he contacted the municipal government about the issue of 'green transportation' and re-construction of the city's bicycle lanes, Chen Jiajun, founder of Biker Guangzhou, met the mayor of Guangzhou and presented him with a bicycle as a gift. This 'gift-giving' caught immediate media attention and won the NGO a valuable window of opportunity to set out its environmental activism work. Since then, Biker Guangzhou has been an active participant in local civil society activities, and shared their experience of public campaign with peer NGOs in Guangdong and beyond. Chen Jiajun has been involved in many peer training programmes and firmly believes in the importance of partnership among grassroots NGOs.[9]

In addition, as the home of the Southern News Group, which publishes the acclaimed *Southern Weekend* and *Southern Cosmopolitan*, Guangdong is arguably the province with the broadest community of 'investigative journalism' and 'critical media' in China. A network of professors, journalists, professionals and 'bloggers' has emerged in Guangzhou since 2005 and started to organize gatherings to promote the idea of 'civic journalism' and train 'barefoot reporters' (individuals who would like to write their blogs on current affairs as journalists) among concerned citizens to publish valuable information via online media outlets (GD Group 1). A new kind of public domain, advanced by creative journalism and public intellectuals in Guangzhou, has brought about important opportunities and vigour to the overall development of civil society in the city and beyond (Yang, 2004; Yang & Calhoun, 2007).

In contrast, civil society in general is underdeveloped in Guangxi, which has direct impact on the current situation of environmental NGOs in the province. The AnDian NPO Centre in Nanning city, the only NGO platform organization in Guangxi, was in fact established with direct guidance from ICS in 2006.[10] With technical support and expertise from NGOs in Guangdong and Hong Kong, plus funding support from Beijing and overseas, AnDian has managed to maintain a work space and a team of two to three full-time staff. It has shared its office with other newly established grassroots NGOs, including the Wild Flower group. Because AnDian still has to struggle with its own capacity-building, its support to other grassroots groups in Guangxi is greatly limited and has not fully developed into the role of ICS as a leading force of social movement and civil society in the province.

Comparing NGOs' vertical and horizontal networks in the two examples illustrates an important dimension of environmental activism at the provincial level. The Guangxi case is more like a vertical extension of environmental activism that originated in Beijing since the 1990s, and overseas contacts and support remain the lifelines to green NGOs' survival. In Guangdong, the environmental activism community has been also influenced by experienced NGOs from outside, yet it increasingly becomes an integral part of the vibrant civil society in the province. Both vertical and horizontal synergies are important to environmental NGOs across the two provinces, but the latter has a strong presence in Guangdong. Among all factors, the leadership of ICS has been crucial to boost the development of civic organizations in Guangdong and the building up of mutual-support networks between them. Lacking such a capable platform organization, Guangxi's civil society sector is still at a nascent stage and dependent on assistance from outside.

Facing the State: Official Attitudes and Policy Adjustments

The existing literature suggests that the state and regulatory settings are the most relevant factors in the development of environmental activism and NGOs in China. This research, therefore, investigates environmentalists and their NGOs' interactions with the state in general and the environmental agencies in particular. First and foremost, most environmentalists think that an accepting and friendly attitude on the part of the state in general has a real—though not direct—effect on their work and organizational development. Second, not all environmental NGOs prioritize networking with governmental agencies in their daily work. NGOs with connections to the authorities resist being dominated by them in subtle yet evident ways. Third, among those NGOs that collaborate with governmental agencies, the propaganda, education and information dissemination agencies affiliated with local environmental protection bureaus (EPBs) are of particular relevance.

Almost all NGO staff and environmentalists who were interviewed agreed on the relevance of the state to the survival of their organization, particularly the official lines and general attitudes shown by key politicians towards NGOs. For example, many mentioned that the China Central Television news coverage of Premier Wen Jiabao shaking hands with a foreign humanitarian NGO representative on his way to meet earthquake victims in Sichuan on 12 May 2008 had a concrete impact in acknowledgeing NGOs' legitimacy in Chinese society. A number of Guangdong environmentalists further commented that they found the public more familiar with the notion of NGOs in recent years due to more official policy initiatives to engage NGOs in governance (GD Group 2). Environmental journalists from Guangdong confirmed that both the provincial- and some local governments had become better educated about what NGOs are and, in turn, more coverage on NGOs appeared in mainstream media (GD Group 1). For example, since 2010, the Shenzhen municipal government has launched propaganda campaigns to advertise its determination to improve public participation in environmental governance, and 'social organization' (*shetuan*) is often mentioned as an important ally.[11] Wang Yang, the governor and party chairman of Guangdong province, has publicly spoken about the importance of 'innovation in social administration' and 'public participation' many times in recent years.[12] He has reportedly taken particular interest in promoting environmental NGOs and the idea of 'democratic environmental protection' (*minzhu huanbao*) since early 2012.[13]

Despite the fact that they think official attitudes and the state matter, not all grassroots NGOs centre their work on collaborating with governmental agencies. Groups such as the Shenzhen Bird-Watching Society and FoN-Shenzhen focus more on their members' concerns. Even when green NGOs need to get governmental agencies 'on board', they make a conscious choice of partnership, in accordance with practical goals and existing personal connections. Dongguan Green Volunteers have a relatively closer relationship with Dongguan municipal EPB compared with other green groups in Guangdong and they acknowledge the EPB's endorsement on their official website. But this NGO still strives for autonomy in designing its educational activities and publishing environmental news.[14] Environmental NGOs are politically conscious—and even savvy—and they bargain with the authorities on a daily basis and in all possible subtle ways (GD Group 2). Newer urban NGOs, exemplified by Biker Guangzhou, sometimes make bolder statements disagreeing with the state by 'symbolic acts'—such as giving the mayor a bike—and offering specific policy recommendations.

The discussions with Guangxi environmental NGOs about governmental relations eventually came round to how hard it was to find sympathetic EPB officials and environmental professionals to work with, and they found local EPBs poorly equipped, understaffed, with little funding to spare and lacking the will to collaborate with NGOs (GX Group). As for Professor Pan and the Dragon Project, they work primarily with village heads and county-level poverty relief, education and public health agencies, and still count on volunteers and experts from outside the province. In a way, the comments from Guangxi environmentalists echo those from their Guangdong peers. Contrary to the common assumption, grassroots NGOs are not only at the mercy of the state. They are aware of, and can assess, local environmental agencies' capacity and professional skills. In a way, they are making the choice between working with and circumventing local authorities. In Guangxi, low governing capacity and the slow pace of reform in policy implementation have generated little incentive for not just local EPBs but also grassroots NGOs to collaborate with each other.

The case of one Guangzhou-based NGO, GP, can further illustrate the practical relationship between local environmental agencies and NGOs when they decide to work with each other. When GP was first established, the group focused on independently conducting environmental public education programmes. They retained a low profile until the Propaganda, Education and Information Centre of the Guangzhou municipal EPB knocked on their door, inviting them to join their work tasks and asking them for advice.[15] Even though this government-affiliated centre was equipped with funding and resources, the officials had little previous experience in working directly with the public and at the community level, such as organizing campaigns at schools and in neighbourhoods. They were looking for collaborators and, having learnt about GP's previous projects, found them to be a most suitable partner. Although the working partnership was not without its difficulties in the following years, both benefited from the collaboration in different ways. GP survived the harsh political environment and became the most experienced environmental NGO in South China, and the Guangzhou EPB learnt many techniques of public education from GP. With higher level of professionalization and organizational capacity, environmental NGOs will have more autonomy in terms whether and how to collaborate with governmental agencies.

What the GP example also implies is that the establishment of new agencies which are mainly designed for managing public participation, access to information and community outreach within the provincial state system can trigger major changes in government–NGO relations on the ground. In recent years, the Guangdong provincial government has made significant progress in engaging the public in environmental governance. According to the Pollution Information Transparency Index (PITI) of China 2010, Guangdong's EPBs performed much better than Guangxi's in terms of disclosure of pollution data, measured by both voluntary information publication and responding to requests (Table 5). Except Nanning, all other Guangxi cities lag behind the Guangdong cities surveyed by PITI and are among the worst of the 113 cities (at prefecture level and above) sampled across the country. The sharp difference in PITI scores between the two cases is partially caused by Guangxi's shortage of institution-building in monitoring, research, propaganda and information dissemination and public outreach (Table 6). It is arguable that there is a rising demand from the government's side in Guangdong to use new methods of reaching out to the public, such as harnessing expertise from environmental NGOs. For the newly established agencies, such as the centres for public environmental education and information dissemination, the need for active NGO participation is even more apparent. The local agencies that are responsible for requests for disclosure of pollution information and complaints will face great challenges in managing com-

Table 6. Numbers of offices under provincial EPB in Guangdong and Guangxi

	General	Enforcement	Monitor	Research	Propaganda	Information centres	Other
Guangdong	21	21	21	18	10	16	40
Guangxi	14	14	15	8	1	4	2

Source: Yang (2010).

munication and interaction with victims. They will be the parts of the state system that are most eager to collaborate with grassroots NGOs in future. The gap between the two provinces in these areas will have major implications for environmental NGOs in the near future.

The analysis presented in this section underlines that government–NGO relationships are no longer one-way or zero-sum. For NGO leaders, official acknowledgement is still a basic requirement for the initial survival of their activism work. However, once an NGO grows its capacity and social networks, it gains opportunity to bargain with the state and can decide whether and how to interact with governmental agencies. When an NGO is specialized in a certain issue area and becomes a leading organization in the field with evident expertise, the state may seek collaboration with it unexpectedly. The Guangdong case indicates that the provincial government's decision to improve public access to environmental information and reform policy-making procedures has provided a new window of opportunity for more grassroots groups to achieve legitimacy and obtain resources, and environmental NGOs such as the GP are prepared for such changes and can cooperate with state agencies without comprising organizational autonomy. The Guangxi case, however, explains that low environmental governing capacity, poor professional development and lack of institutional arrangements for public participation constrain the long-term trajectory of the NGO sector, although they do not threaten the existence of NGOs as much as political hostility.

Conclusions

Environmental activism and policy advocacy in China is not a single national phenomenon but an amalgamation of grassroots initiatives essentially led by environmental NGOs, which has gradually taken root in provincial and local politics. This paper has examined the current development of environmental NGOs and activism in Guangdong and Guangxi, and highlighted the impact of inter-organizational relations. The first and most important implication of the study is the rich, compound and diverse nature of environmental activism and NGO communities in China's provinces. The green NGO sectors in the two provinces differ in many respects, not just in numbers and scope. Although all of the grassroots NGOs studied in this research are related to the overall rise of environmental activism originated in Beijing in the 1990s to different degrees, Guangdong environmental NGOs show a stronger integration within the civil society sector in their home province.

Second, each green activism community exists in a complex socio-political environment. Different levels of development in economy, media and higher education condition the emergence of environmental activism and NGOs, and to some extent explain the trends of post-materialistic values exhibited by a few NGOs in the cities of Guangdong. The presence of a leading platform organization that makes a specific effort to promote the idea of 'civil society' and nurture NGO networks stands out as a major factor for the steady growth of environmental activism in Guangdong. The most important life-saving assistance for many of the NGOs in Guangdong clearly comes no longer from the state but from peer organizations. Clusters of NGOs and platform organizations have gradually generated resources within local civil society realms, independent from the state system. To a degree, the level of horizontal power generated by civic organization determines whether the initial growth of environmental activism at the local level can stabilize and eventually blossom.

Last but not least, findings on how governmental and societal structures have affected environmental NGOs in Guangdong and Guangxi challenge the pessimistic view that NGOs are reactive agents and inevitably trapped in the ubiquitous forms of state domination in China. The state matters, but not always, or uniformly. The emerging NGO sector has built up its 'bargaining muscles' by extending both vertical and horizontal networks. This research has found that many NGO leaders are politically conscious and capable of making informed decisions when interacting with government agencies. They accept official invitations when the opportunity is right. As the Chinese government is taking more substantial steps in social reform and enhancement of public participation exemplified by the policy adjustments in Guangdong, the demand of NGO partnership will rise and the existing dynamics between the state and the NGO sector will shift further. Having taken an active role in strengthening civil society in the province, Guangdong environmental groups are more likely to establish broad coalitions, launch policy advocacy campaigns and together negotiate with the state. However, in Guangxi and the majority of Chinese provinces, environmental NGOs are scattered among local communities. Despite international support and acceptance by the local governments, environmental activism with low levels of horizontal synergies and peer connections will remain fragmented and vulnerable to state interference.

Notes

1. There is so far a clear disconnection in practice between the two kinds of citizens' actions in contemporary China. Protests by rural farmers or urban residents affected by industrial pollution or large-scale infrastructure construction broke out in China as early as the 1980s and became common in the 1990s (Jing, 2000). They can be better explained by the concept of 'rightful resistance' coined by O'Brien & Li (2006) rather than so-called 'NIMBYism' (Not-In-My-Back-Yard). In recent prototype NIMBY campaigns, as shown by the detailed studies of Johnson (2010) and Van Rooij (2010), the activist community has been cautious and made conscious decisions to avoid being directly or fully involved. Also, the campaigners emerging from residential communities, who may have succeeded in protecting their own interests and neighbourhood, usually do not turn into committed environmentalists to pursue long-term or policy changes (Stern, 2008; Zhu & Ho, 2008).
2. These surveys were conducted independently by three NGOs, FoN—the most recognized environmental NGO in China, GSEAN (*Jixi*)—a nationwide network of university-based groups and NGOCN—a web-based non-profit enterprise serving the entire community of civil society.
3. Hu Jintao's speech at the 17th Congress of Chinese Communist Party, 24 October 2007. Available at http://news.xinhuanet.com/newscenter/2007-10/24/content_6938568.htm (accessed 22 November 2012).
4. Guo Hong, director of the 5.12 Non-Governmental Relief Service Centre in Chengdu city, which coordinated all NGO efforts in helping the affected communities in Sichuan province since 2008, concurred with this observation about the environmental NGOs when interviewed on 28 April 2012.
5. Two of these awards are sponsored by the Ministry of Environmental Protection, i.e. the 'China Environmental Protection Award' and the 'Green Chinese Award' (*lvse zhongguo niandu renwu*) (2005–2009). The others include the China Earth Award (1997–2003) co-sponsored by the Hong Kong environmental NGO Friends of the Earth and the China National Association of Journalists, the Ford Motor Environmental Award (2000–2008), sponsored by Ford Motor (China) Ltd. and the SEE TNC Award (2007–2009), sponsored by the Beijing-based NGO, the Alxa Society of Entrepre-

neurs and Ecology (a.k.a. *A La Shan*) and the China office of an American NGO, The Nature Conservancy. The latter three clearly give greater priority to non-state-affiliated individuals and organizations.

6. Interview in Nanning, 10 May 2008.
7. Interviews in Nanning on 10 and 12 May 2008 and follow-up phone interviews since then, until March 2011.
8. Office visits in Guangzhou on 18 July 2008 and 13 March 2012.
9. Interview in Hong Kong in December 2011.
10. Interview in Nanning, 3 May 2008.
11. The Shenzhen government launched an official website called 'Living Environment' (*renju huanjing wang*) to encourage the public to process requests, access official information and submit feedback: http://www.szhec.gov.cn.
12. In his most recent official visit to Singapore in June 2012, 'social innovation' and 'public participation' were among the priority tasks for discussion and learning. (http://zt.my/News/NewsShow-42054.html)
13. Interviews with Mr Huang, the executive director of a Guangzhou-based private charity, and Mr Zhang, the founding director of the Green Point in Hong Kong, May 2012. The 2012 August issue of the *China Youth Environmental Review*, an independent publication widely acclaimed among environmentalists, published Wang Yang's idea of *minzhu huanbao* in detail. (http://blog.sina.com.cn/s/articlelist_1882201622_3_1.html)
14. Phone interview, 25 July 2011.
15. Interview in Guangzhou, 17 July 2008.

References

Baldassarri, D. & Diani, M. (2007) The integrative power of civic networks, *American Journal of Sociology*, 113(3), pp. 735–780.

Buesgen, M. (2008) Environmental NGOs' role in expanding social spaces-diversification with Chinese characteristics: A case study of NGOs' opposition to the Nujiang Dam in China's Yunnan province, *China Journal of Social Work*, 1(2), pp. 160–171.

Carroll, W. K. & Ratner, R. S. (1996) Master framing and cross-movement networking in contemporary social movements, *Sociological Quarterly*, 37, pp. 601–625.

Carter, N. T. & Mol, A. P. J. (Eds) (2007) *Environmental Governance in China* (London: Routledge).

Chen, J. (2010) Transnational environmental movement: Impacts on the green civil society in China, *Journal of Contemporary China*, 19(65), pp. 503–523.

Cooper, C. (2006) "This is our way in": The civil society of environmental NGOs in south-west China, *Government & Opposition*, 41(1), pp. 109–136.

Deng, G. (2010) The hidden rules governing China's unregistered NGOs: Management and consequences, *China Review*, 10(1), pp. 183–206.

Directory of Environmental NGOs (2005) *Green Book of China's Environment* (Friends of Nature and Yang Dongping, Eds) (Beijing: China Social Science Academy Press).

Economy, E. (2004) *The River Runs Black: The Environmental Challenge to China's Future*, pp. 129–176 (Ithaca, NY: Cornell University Press).

Goldfarb, J. C. (2001) 1989 and the creativity of the political, *Social Research*, 68(4), pp. 993–1010.

Hayes, J. P. (2007) The recent environmental history of tiger leaping gorge: Environmental degradation and local land development in northern Yunnan, *Journal of Contemporary China*, 16(52), pp. 499–516.

Hildebrandt, T. (2011) The political economy of social organization registration in China, *The China Quarterly*, 208, pp. 970–989.

Ho, P. & Edmonds, R. L. (Eds) (2008) *China's Embedded Activism: Opportunities and Constraints of a Social Movement* (New York: Routledge).

Ho, P. & Vermeer, E. B. (Eds) (2006) Special volume on "China's Limits to Growth: Greening State and Society", *Development and Change*, 37(1).

Inglehart, R. (1977) *The Silent Revolution: Changing Values and Political Styles among Western Publics* (Princeton, NJ: Princeton University Press).

Jing, J. (2000) Environmental protests in rural China, in: E. Perry & M. Selden (Eds), *Chinese Society: Change, Conflict and Resistance*, pp. 143–160 (New York: Routledge).

Johnson, T. (2010) Environmentalism and NIMBYism in China: Promoting a rules-based approach to public participation, *Environmental Politics*, 19(3), pp. 430–448.

Kubik, J. (1994) *The Power of Symbols against the Symbols of Power: The Rise of Solidarity and the Fall of State Socialism in Poland* (University Park, PA: Pennsylvania State University Press).

Map of Chinese Environmental NGOs (2009). GSEAN. Available at http://www.gsean.org (accessed 25 August 2010).

Matynia, E. (2001) The lost treasure of solidarity, *Social Research*, 68(4), pp. 917–936.

Mertha, A. (2008) *China's Water Warriors: Citizen Action and Policy Change* (Ithaca, NY: Cornell University Press).

Mol, A. P. J. (2006) Environment and modernity in transitional China: Frontiers of ecological modernization, *Development & Change*, 37(1), pp. 29–56.

Mol, A. P. J. & Carter, N. T. (2006) China's environmental governance in transition, *Environmental Politics*, 15(2), pp. 149–170.

Morton, K. (2005) The emergence of NGOs in China and their transnational linkages: Implications for domestic reform, *Australian Journal of International Affairs*, 59(4), pp. 519–532.

NGOCN Online Database of Environmental NGOs (2010). Available at http://ngocn.net (accessed 25 August 2010).

Obach, B. K. (2004) *Labor and the Environmental Movement: The Quest for Common Ground* (Cambridge, MA: MIT Press).

O'Brien, K. & Li, L. (2006) *Rightful Resistance in Rural China* (Cambridge: Cambridge University Press).

Rochon, T. R. & Meyer, D. S. (1997) *Coalitions & Political Movements: The Lessons of the Nuclear Freeze* (Boulder, CO: Lynn Rienner).

Ru, J. & Ortolano, L. (2009) Development of citizen-organized environmental NGOs in China, *Voluntas: International Journal of Voluntary & Nonprofit Organizations*, 20(2), pp. 141–168.

Saich, T. (2000) Negotiating the state: The development of social organizations in China, *The China Quarterly*, 161, pp. 125–141.

Salamon, L. & Anheier, H. K. (1997) *Defining the Nonprofit Sector* (Manchester: Manchester University Press).

Shieh, S. & Deng, G. (2011) An emerging civil society: The impact of the 2008 Sichuan earthquake on grass-roots associations in China, *The China Journal*, 65, pp. 181–194.

Spires, A. J. (2007) China's un-official civil society: The development of grassroots NGOs in an authoritarian state, PhD dissertation, New Haven, Yale University.

Spires, A. J. (2011) Contingent symbiosis and civil society in an authoritarian state: Understanding the survival of China's grassroots NGOs, *American Journal of Sociology*, 117(1), pp. 1–45.

Stern, R. (2008) From rhetoric to action: Talking about environmental rights in contemporary China, Paper presented at the ISA's 49th Annual Convention, San Francisco, CA, USA, March 26.

Sullivan, J. & Xie, L. (2009) Environmental activism, *Social Networks and the Internet, China Quarterly*, 198, pp. 422–432.

Sun, Y. & Zhao, D. (2008) State-society relations and environmental campaigns, in: K. O'Brien (Eds), *Popular Protest in China*, pp. 145–162 (Cambridge, MA: Harvard University Press).

Tang, S. N. (2008) Civic environmental NGOs, civil society, and democratization in China, *The Journal of Development Studies*, 44(3), pp. 425–448.

Tismaneanu, V. (Ed.) (1990) *In Search for Civil Society: Independent Peace Movement in the Soviet Bloc* (New York: Routledge).

Tismaneanu, V. (1993) *Reinventing Politics: Eastern Europe from Stalin to Havel* (New York: The Free Press).

Turner, J. (2004) Small government, big and green society: Emerging partnership to solve China's environmental problems, *Harvard Asia Quarterly*, VIII(2), Available at http://www.asiaquarterly.com (accessed 25 March 2005).

Turner, J. & Wu, F. (Eds) (2001) *Green NGO and Environmental Journalist Forum: A Meeting of Environmentalists from Mainland China, Taiwan and Hong Kong* (Washington, DC: Woodrow Wilson International Center for Scholars).

Van Dyke, N. & Mccammon, H. J. (2010) *Strategic Alliances: Coalition Building and Social Movements* (Minneapolis, MN: University of Minnesota Press).

Van Rooij, B. (2010) The people vs. pollution: Understanding citizen action against pollution in China, *Journal of Contemporary China*, 19(63), pp. 55–77.

Weiss, M. L. (2006) *Protest and Possibilities: Civil Society and Coalitions for Political Change in Malaysia* (Stanford, CA: Stanford University Press).

Wu, F. (2009) *Environmental Activism in China: Fifteen Years in Review, 1994–2008*. Harvard-Yenching Working Paper Series. Available at http://www.harvard-yenching.org/features/hyi-working-paper-series-wu-fengshi (accessed 1 August 2011).
Wu, F. & Chan, K. M. (2012) Graduated control and beyond: The evolving governance over social organizations in China, *China Perspectives*, 3, pp. 9–17.
Xie, L. (2009) *Environmental Activism in China* (New York: Routledge).
Xie, L. (2011) China's environmental activism in the age of globalization, *Asian Politics and Policy*, 3(2), pp. 207–224.
Yang, G. (2004) *Global Environmentalism Hits China: International and Domestic Groups Join Forces to Combat Environmental Woes*. Working paper at the Yale Center for the Study of Globalization. Available at http://yaleglobal.yale.edu/display.article?id=3250 (accessed 25 March 2005).
Yang, G. (2005) Environmental NGOs and institutional dynamics in China, *The China Quarterly*, 182, pp. 46–66.
Yang, G. & Calhoun, G. (2007) Media, civil society, and the rise of a green public sphere in China, *China Information*, XXI(2), pp. 211–236.
Yang, M. (Ed.) (2010) *China Environmental Yearbook*. Beijing: China Environmental Yearbook Publishing House.
Yang, X. & Gabriel, G. (2001) NGO Conservation Work in Tibet and Western Sichuan. Paper presented at the China Environment Forum Meeting, Woodrow Wilson International Center for Scholars, Washington, DC. Available at http://www.wilsoncenter.org (accessed 28 August 2001).
Zhu, J. & Ho, P. (2008) Not against the state, just protecting residents' interests: An urban movement in a Shanghai neighborhood, in: P. Ho & R. Edmonds (Eds), *China's Embedded Activism*, pp. 151–170 (New York: Routledge).

Appendix. Focussed group discussion with environmental, related NGOs and journalists in Guangdong and Guangxi.

Guangdong

(GD Group 1) Guangzhou, 16 June 2008. Two investigative journalists for the Southern News Group and two researchers, members of the 'Civic Journalism Forum', who wish to remain anonymous.

(GD Group 2) Guangzhou, 17 July 2008. One staff member from the GP and one staff member from the ICS, Sun Yat-Sen University.

Guangxi

(GX Group) Nanning, 12 May 2008. Two participants from the Green Salon of the Guangxi Medical University and one staff member from Green Alliance of Guangxi.

The Politics of Waste Incineration in Beijing: The Limits of a Top-Down Approach?

THOMAS JOHNSON

Department of Public and Social Administration, City University of Hong Kong, Tat Chee Avenue, Kowloon, Hong Kong SAR

ABSTRACT *The siting of waste incinerators is often a highly contentious issue. Although most studies have focused on Western countries, municipal solid waste management has become an extremely pressing issue in China. Incineration is being strongly promoted by government officials, yet this has resulted in strong societal opposition. Through documentary analysis and in-depth stakeholder interviews, this article analyses three anti-incineration campaigns in Beijing, focusing on outcomes and campaigner strategies. Anti-incineration campaigns have partly undermined the government's top-down, non-consultative approach to waste management. In developing an 'expert strategy', campaigners have exploited government weakness whilst depoliticizing the issue. Yet rather than lead to a more open and consultative incineration policy, it is more likely that officials will circumvent unrest through increasing opacity and by choosing sites in locations where opposition is less likely to emerge.*

Introduction

Increasing quantities of municipal solid waste (MSW) are presenting a growing challenge for urban governments across the world. Far from being merely a technical process, MSW management is often highly contentious. The issue of waste incineration has been particularly controversial due to public concerns surrounding public health and the environment (Rootes, 2009). Opponents claim that waste incineration emits dangerous, polluting, and highly carcinogenic substances including dioxins, mercury, and furans, and undermines more sustainable waste treatment methods such as recycling and waste reduction. Conversely, proponents point to benefits such as reduced pressure on landfill sites and energy generation as a by-product of waste-to-energy (WTE) incinerators, whilst insisting that emissions can be controlled within strict standards. Issues surrounding waste incineration have caused numerous collisions of interest between different levels of government, corporations, environmental non-governmental organizations (NGOs), and communities opposing decisions to site MSW facilities in their 'backyards'. Such conflicts sometimes facilitate constructive discussion and debate concerning waste management solutions (McAvoy, 1999; Nissim *et al.*, 2005). At the same time, prolonged conflict can undermine waste treatment, resulting in environmental degradation and health risks (Davies, 2008).

Scholarly understandings of MSW management controversies are almost exclusively confined to cases from advanced industrialized democracies (see, for example, Davies, 2008; Rootes & Leonard, 2010; Walsh *et al.*, 1997). And yet MSW management is emerging as a particularly pressing and contentious issue in emerging economies. No where we face a greater challenge in this respect than China, which at the start of the millennium overtook the USA to become the world's largest generator of MSW (World Bank, 2005). As of 2009, China was producing approximately 148 million tonnes of MSW annually (Asian Development Bank, 2009), a figure that was forecast to more than treble to 480 million tonnes per annum by 2030 (World Bank, 2005). The twin processes of unprecedented rapid urbanization and changing consumption patterns are the main drivers behind this exponential growth. By 2008 China's urban population stood at over 600 million compared with 190 million in 1980 (*China Daily*, 2009). It is estimated that by 2050 Chinese urban dwellers, who on average generate between two and three times more waste than their rural counterparts, will represent 75% of the population (Wu *et al.*, 2007). This is placing intense strain on the country's underdeveloped waste treatment infrastructure. According to the Asian Development Bank (2009), almost half of China's MSW is untreated and disposed of in unsuitable landfills on the outskirts of cities. The OECD (2007, p. 131) has claimed that aerial surveys show many urban areas 'are almost surrounded by dumps'. On 1 April 2009, the state-owned *People's Daily* reported that one-third of Chinese cities were surrounded by waste, covering a total area of 500 million square metres (*People's Daily*, 2009).

In common with many advanced industrialized nations in the early 1970s, the Chinese central government has embraced waste incineration in response to a considerable MSW challenge. The central government aims to increase the amount of waste disposed of in WTE incinerators from 1% in 2002 to 30% by 2030, necessitating a massive expansion in China's waste incineration capacity (Asian Development Bank, 2009). At the end of 2004, China only had 54 municipal waste incinerators (OECD, 2007). By 2010 this had almost doubled to 103 incinerators, and it is expected that China will have over 300 incinerators with a combined daily capacity of 300,000 tonnes by the end of the current 12th Five-Year Plan (FYP) (2011–2015) (*New Century*, 2012).

This rapid expansion in waste incineration capacity has incited opposition campaigns from local communities. According to *The Beijing News*, between 2008 and 2009 'anti-incinerator-construction incidents' occurred in 30 Chinese cities (*The Beijing News*, 2010c). This has potentially serious implications both for China's waste treatment strategy and the Chinese Communist Party's (CCP) core goal of maintaining social stability. Waste incineration opposition has been particularly prominent in Beijing, where the Municipal Government announced plans to construct nine incinerators with a combined daily capacity of 8200 ton by 2015 (Yang, 2010). Since 2006, residents in suburbs of Liulitun, Gaoantun, and Asuwei, located in the Haidian, Chaoyang, and Changping districts, respectively, have waged high-profile campaigns against decisions to site incinerators in their neighbourhoods. Although these three areas fall under the same municipal government jurisdiction, campaign outcomes and strategies varied widely. This article has two main aims. First, it examines how citizens in these three cases mobilized against incinerators, and analyses the factors contributing to campaign outcomes. Second, it considers the government's response to anti-incineration campaigns and the broader implications for waste governance in Beijing and China more generally. The next section of this article briefly examines how anti-siting movements emerged and developed in a Western context, before providing an overview of Beijing's 'waste crisis'. It then examines the three anti-incineration campaigns in detail and discusses government responses before finally drawing some conclusions.

Anti-Siting Movements in a Comparative Perspective

During the late 1960s and early 1970s, citizen's concern regarding linkages between industrialism and public health in advanced industrialized nations increased (Rootes, 2009). This concern was partly manifested in growing localized opposition to the siting of 'locally unwanted land uses' (LULUs), including waste incinerators. Campaigners have used an array of strategies to oppose LULU sitings, including lobbying, litigation, and extra-institutional tactics (Rootes & Leonard, 2010; Walsh *et al.*, 1997).

When anti-siting campaigns first emerged, project proponents frequently labelled campaigners as troublesome NIMBY ('not-in-my-backyard') actors who selfishly put their own narrow concerns ahead of the public good. NIMBY actors were portrayed as overly emotional, lacking in technical expertise, and motivated by irrational concerns (McAvoy, 1998). Partly as a result, government officials were (and still are) largely unwilling to incorporate public opinion into siting decisions (McAvoy, 1998). Although attempts have been made to solicit the opinions of local communities, including via public hearings and, more recently, through deliberative processes such as citizens' advisory committees (Petts, 2001), these efforts have often resulted in unsatisfactory levels of public participation. In some cases, public hearings have been viewed as tools of public appeasement with minimal impact on decisions (Joppke, 1993). Moreover, low levels of transparency, including the apparent deliberate under-publicizing of public hearings, have reinforced public scepticism (Cole & Foster, 2001; Shevory, 2007; Walsh *et al.*, 1997). This has sometimes resulted in elevated levels of anger and mistrust directed against project proponents (Innes & Booher, 2004).

Growing opposition to LULUs such as waste facilities meant that political factors became more important in siting decisions. In some cases, governments and corporations strategically located waste facilities in or close to communities that were, based on their demographic characteristics, predicted to offer least resistance. When a 1984 report identifying characteristics of communities supposedly least hostile to waste incinerator siting that had been commissioned by the California Waste Management Board found its way into activists' hands, this strategy was laid bare for all to see (Cole & Foster, 2001; Walsh *et al.*, 1997). Evidence that a high number of toxic dumps were sited in ethnic minority communities in the USA fuelled the environmental justice movement (Cole & Foster, 2001). Atlas (2001a, 2001b) found that by the 1990s landfill expansion projects were less likely to proceed in areas with high concentrations of ethnic minorities, and that minority and low-income communities did not bear a disproportionate risk from hazardous waste facilities. According to Rootes and Leonard (2009), this was testament to the environmental justice movement's success.

Indeed, the networking of disparate campaigns into broader social movements is a striking feature of anti-siting campaigns in the West. Anti-siting campaigns first emerged in isolation from each other and from a broader environmental agenda, with early victories against LULUs reflecting the strength of local opposition groups rather than that of a wider movement (Gottlieb, 1993). In the USA, a myriad of grassroots organizations established to oppose unwanted waste facilities formed links with the wider environmental justice movement (Fletcher, 2003; Rootes & Leonard, 2009). As well as providing local activists with information and support and making it harder for facility proponents to discredit opponents, this has been credited with rejuvenating the US environmental movement (Rootes & Leonard, 2009; Walsh *et al.*, 1997). Although the environmental justice discourse has been far less prominent in Europe, campaigners have also, to varying degrees, networked into movements. This has sometimes been achieved with the support of established national-level environmental NGOs, such as in England (Rootes, 2009). The networking of anti-siting campaigns has been seen as a key factor in diverting attention away from parochial NIMBY sentiment towards a wider concern about the relationship between industrialism, health, and environment at a regional, national, and global level (Freudenberg & Steinsapir, 1992; Hess, 2007; Rootes & Leonard, 2009; Shemtov, 1999). In the early 1970s, governments across the USA viewed waste incineration as an effective policy solution, yet by the mid-1980s public opposition had made it extremely difficult to build an incinerator (Walsh *et al.*, 1997). Rootes and Leonard (2009) argue that anti-incineration campaigns were an important factor in pushing the American waste industry towards a more sustainable direction. Even when unsuccessful, campaigners have helped to promote public participation and information disclosure (Gottlieb, 1993; Rootes & Leonard, 2009).

In China, anti-siting campaigns have emerged as part of a wider upsurge in localized, single-issue contention since the reform and opening-up period began in 1978 (Perry & Selden, 2003). Influenced by the CCP's official commitment to rule of law, increasingly rights-conscious citizens tend to utilize 'rights upholding' (*weiquan*) language, whereby they call on unruly officials to implement policies and govern according to laws and regulations. Since 2003, the central government has provided formal channels for public consultation in laws and measures pertaining to environmental impact assessment (EIA) and administrative licensing, and has also passed measures requiring greater environmental information disclosure (Johnson, 2010). Although this legislation has numerous shortcomings and has only been weakly enforced, ordinary citizens are increasingly aware of their 'right' to be consulted over the siting of waste incinerators and other large facilities (Johnson, 2010).

Incineration Policy and Public Opposition in Beijing

In the early 1980s, over 4700 piles of waste that exceeded 50 square metres were identified between Beijing's third and fourth ring roads. In response, the Beijing Municipal Government invested 2.3 billion Yuan in constructing 23 waste facilities. Then, landfill was the preferred solution to Beijing's waste problem. However, a sharp increase in the generation of waste coupled with the rapid filling up of landfills has created a significant challenge for municipal authorities. On 3 September 2009, head of the Beijing Municipal Commission of City Administration and Environment (BMCCAE), Chen Yong, publicly stated that the city was facing a 'waste crisis'. Current waste treatment capacity was just 10,300 ton per day, far below the 18,400 ton of waste that were being generated daily. The Beijing Municipal Government responded by announcing plans to massively expand the city's daily waste treatment capacity to 17,000 and 30,000 ton by 2012 and 2030, respectively (Chen, 2010).

In order to meet its waste treatment target, the Beijing Municipal Government aims to significantly increase the proportion of incinerated waste. In 2007, it published the 'Beijing Municipality Solid Waste Treatment Plan for the 11th Five-Year Plan Period'. Pursuant to this, the Municipal Development and Reform Commission (DRC) published the 'Implementing Plan for Beijing Municipality Domestic Waste Treatment Facility Construction in the 11th Five-Year Plan Period' (hereinafter referred to as the 'DRC Domestic Waste Plan') (Beijing DRC, 2007). Incineration was central to this document, with pollution concerns, land scarcity, and cost cited as reasons against expanding landfill (Beijing DRC, 2007). At the start of the Plan period in 2006, government statistics stated that 90% of Beijing's treated waste went to landfill, with 8% and 2% being composted and incinerated, respectively.[1] According to the Beijing DRC, by the end of the Plan period in 2010, 40% of waste in the city area would be incinerated, with landfill and composting both accounting for 30% of waste treatment (Beijing DRC, 2007).[2] This was labelled the '4:3:3' objective with reference to the proposed ratio between incineration, landfill, and composting.[3] In comparison, 5.4% of European Union (EU) waste and 12.7% of the US waste were incinerated in 2008 (Eurostat, 2010; *The New York Times*, 2008); in Japan, almost 80% of waste was being incinerated as of 2005 (*The New York Times*, 2005). To meet the 4:3:3 objective, four incinerators were to be constructed and/or go into operation during the Plan period at Liulitun, Gaoantun, Asuwei, and Nangong (Beijing DRC, 2007). This section examines the first three cases. Nangong was not chosen, partly due to space constraints, and also because societal opposition was weak.

Liulitun: 'Seeking Truth from Facts'

In 1995, approval was given to construct a landfill facility at Liulitun, located in Northwest Beijing's Haidian district. Yet from the beginning, there was disagreement within government regarding the suitability of the site, which was chosen due to the existence of a pit created by a brick-firing company. The Beijing Municipal Environmental Protection Bureau (EPB)'s EIA report stated that, 'from an environmental protection viewpoint, it is unsuitable to construct a waste landfill at this location' (Feng, 2009, p. 31). Although it subsequently agreed to the landfill, the EPB suggested that no housing be built within 500 m of the site, and that local residents be relocated. These suggestions were ignored. Moreover, the surrounding area continued to be developed for residential purposes.

After the first phase of the landfill project was completed in 1999, local residents started complaining about a stench from the site. Subsequently, concerns were voiced that the landfill had contaminated groundwater, causing health problems among local residents, including increasing cases of cancer (Liu, 2007). In late 2006, residents were, therefore, horrified to learn of plans to build an incinerator capable of burning 1200 ton of waste per day adjacent to the landfill, which was due to be decommissioned in 2012.[4] The Municipal EPB had approved the incinerator project's EIA in November 2005, with construction scheduled to start in March 2007 to ensure completion before the 2008 Beijing Olympic Games.

Residents wasted little time in mobilizing against the incinerator. They utilized *xinfang* and *shangfang* ('letters and visits') channels, which involve petitioning government departments in person, through letters and emails, and via phone calls. Campaigners focused on what they claimed was an arbitrary, or 'unscientific' decision-making process that had overlooked the unsuitability of the Liulitun site, and which contravened regulations governing where incinerators can be cited.[5] Local residents formed a 'study group' composed of several retirees to conduct research into local geographic conditions, as well as policy and technical issues related to incineration. Campaigners also visited potentially sympathetic experts to obtain technical and legal information and were encouraged by support received from several of them (Feng, 2010). According to the 2006 State Environmental Protection Administration (SEPA) and National Development and Reform Commission (NDRC) 'Notice Regarding Strengthening of EIA Management Work for Projects Involving Electricity Generation from Biomass' (SEPA & NDRC, 2006), incinerators 'ordinarily' should not be built in the following locations:

- Built-up or planned urban areas in large- and medium-sized cities.
- Upwind from towns and cities, and/or densely populated areas.
- Environmentally sensitive areas where an incinerator might result in failure to meet environmental targets and standards.

Local residents argued that Liulitun had become a densely populated area, with over 250,000 people living within 5 km of the site. They claimed that Liulitun was an environmentally sensitive area, with the Miyun-to-Beijing water channel that supplies Beijing's drinking water only 1 km from the proposed incinerator. Furthermore, the local area was environmentally sensitive due to pollution from the landfill and the fact that many historical sites and universities were situated nearby. In January 2007, campaigners obtained an abridged version of the incinerator project's EIA report from the Beijing EPB website, and immediately found mistakes and inconsistencies.[6] For example, residents questioned the report's claims that dioxins can be measured in real-time, and argued that the prevailing wind in Beijing is north-westerly (which would place Liulitun upwind from downtown Beijing) and not, as the EIA claimed, north-easterly. Residents also complained that they had not been consulted about the project.

After receiving advice from environmental lawyer Xia Jun, in February 2007 campaigners applied to SEPA for an administrative review into the Beijing EPB's 'illegal' decision to approve the project's EIA. They asked SEPA to order the municipal EPB to revoke the original EIA, conduct a new EIA incorporating public consultation, and to temporarily suspend the project (SEPA, 2007b).

According to the Administrative Review Law, government departments should respond within 90 days. Having received no response, and frustrated with the lack of progress, over 1000 residents converged on SEPA's headquarters in Beijing demanding action from the government on 5 June 2007 (World Environment Day).[7] This proved to be an important turning point in the case.[8] SEPA Vice-Minister Pan Yue was given responsibility for overseeing the Liulitun case (Interview with environmental lawyer, Beijing, 29 November 2010). Campaigners viewed this as a fortunate development (Interview with Liulitun campaigners,

Beijing, 6 June 2011). Between 2005 and 2007, Pan had launched three high-profile 'EIA storm' campaigns designed to improve implementation of the EIA Law. Also around this time, Pan was particularly determined to promote greater public participation and transparency in the EIA process. According to a lawyer with close links to SEPA, Pan wanted to use the Liulitun case to further his agenda of improving public participation in EIA (Interview with environmental lawyer, Beijing, 29 November 2010). In a statement issued two days after the protest, Pan stated that, 'we hope that all levels of government will provide a platform to enable the public to enjoy sufficient right to know, right to supervise, and right to participate in large-scale environmental affairs' (SEPA, 2007a). The same statement suggested that the project be suspended to allow further discussion and soliciting of public opinion (SEPA, 2007a). The administrative review decision, issued on 12 June, echoed this sentiment. It criticized local officials for ignoring the EPB's 1995 advice and allowing the area surrounding the landfill to be developed for residential purposes. However, it absolved the EPB of any wrongdoing and upheld the legality of the EIA (SEPA, 2007b). Nevertheless, SEPA reiterated its recommendation that the project be temporarily suspended pending further inquiries (SEPA, 2007b).

SEPA's intervention did not quell the strong support for the incinerator within the Municipal and Haidian District governments. Nevertheless, construction plans had apparently been suspended. As campaigners were fully aware, the rapidly approaching 2008 Beijing Olympics ensured that the government did not want to provoke further social unrest. In the meantime, residents continued petitioning the government through letters and visits, although they consciously tried to avoid making the government 'nervous' by applying too much pressure (Interview with Liulitun campaigners, Beijing, 6 June 2011). Then, in early 2009, the Beijing Municipal DRC named the Liulitun incinerator among a list of major construction projects for that year. Residents responded with a petition letter containing over 10,000 signatures. The letter expressed frustration and anger that the Haidian District Government had ignored SEPA's instructions to solicit public opinion, and reiterated their concerns regarding public health. On 31 March 2009, residents produced a detailed 40,000-character report entitled 'Opinion Letter for Opposing the Construction of a Waste Incinerator in Liulitun' (hereinafter referred to as 'Opinion Letter') that they sent to numerous Beijing and national government departments (Beijing Haidian District Residents and Enterprise Staff, 2009). As well as further explicating earlier concerns, this report presented several suggestions to the government regarding its waste policy. In contrast to earlier in the campaign, residents portrayed themselves as trying to help the government resolve the waste issue. However, they maintained their strong opposition to the 'irresponsible' siting decision and implored the government to decide rationally and 'seek truth from facts'. As residents pointed out in their 'Opinion Letter', several influential public figures agreed that the siting decision was flawed. For example, in 2007, a member of the China People's Political Consultative Conference (CPPCC) had filed a report proposing that the incinerator be cancelled, as did a member of the Haidian District CPPCC. Qu Geping, the former head of the national environmental protection agency, publicly voiced his opposition to the choice of location. The decision to abandon construction of the Liulitun incinerator emerged in drips and drabs. In late 2010, the Haidian District Government announced plans to construct an incinerator, apparently in lieu of the Liulitun incinerator, at the much less densely populated Sujiatuo, approximately

20 km from Liulitun. Finally, in January 2011 and over 4 years since the project was announced, officials confirmed that the Liulitun incinerator would not be built.

Gaoantun: 'The Rice Has Been Cooked'

In 1995, a rubbish dump at Gaoantun in Beijing's Chaoyang District was converted into a waste treatment plant, and in 2002 a landfill was opened at the site. Similar to Liulitun, residents noticed a stench from the landfill that became progressively worse. The surrounding area was also developed for residential purposes. In 2004, local residents began petitioning the government via *shangfang* and *xinfang* channels, and filed a petition of 240 signatures. Officials promised them that action was being taken over the stench, but the problem remained unresolved (Interview with Gaoantun campaigner, Beijing, 6 June 2011).

Around the same time, and unbeknownst to residents, government officials had chosen Gaoantun as an incinerator site. In 1999, SEPA approved an EIA for an incinerator to be built next to the landfill. However, construction was delayed for several years, necessitating a second 'adjustment' (*tiaozheng*) EIA in 2004 that SEPA also approved. Construction of the Gaoantun incinerator started in November 2005, yet residents did not mobilize against it until 2007, by which time construction was virtually completed. One resident had noticed the incinerator but thought it was a factory (Interview with Gaoantun campaigner, Beijing, 6 June 2011). At that time, residents' knowledge of waste incineration was virtually non-existent.

The Liulitun case, which attracted considerable media attention, strongly influenced Gaoantun residents. Campaigners were alerted to the 'dangers of incineration' by the World Environment Day protest (Home Defence Action Group, 2008). Having become aware of the Gaoantun incinerator, several residents from the middle-class Wanxiang Xintian and Bolin Aile housing communities formed the Home Defence Action Group (HDAG) to oppose it. Similar to Liulitun campaigners, Gaoantun residents focused their opposition on legal issues related to the incinerator siting decision. Residents faxed SEPA in July and August 2007 asking to see the incinerator's EIA report. After SEPA refused, the HDAG held a meeting of about 40 residents in September where it revealed plans to submit an administrative review application challenging the decision to approve the EIA. HDAG organizers felt compelled to resort to a campaign of 'protecting our home through legal weapons' because government officials had not engaged them or addressed their concerns. Residents raised money and enlisted the services of Xia Jun following his involvement in the Liulitun case.

The residents filed their administrative review application with SEPA in October 2007. It requested that the 2004 EIA be revoked and that the incinerator be either terminated or temporarily suspended. When preparing the administrative review application, residents were allowed to see the abridged EIA but not to photocopy it. Similar to Liulitun campaigners, the administrative review application questioned the site's suitability and also criticized a lack of due process in the handling of the EIA (Zhou *et al.*, 2007). In relation to the former issue, residents argued that Gaoantun was an *even less* suitable location for an incinerator than Liulitun, which was, at this stage, on hold following SEPA's intervention.[9] They pointed out that the incinerator was upwind from densely populated areas. Residents also claimed that the 2004 adjustment EIA had not considered changes in local land use occurring since the original EIA was conducted in

1999. In that time, and similar to Liulitun, the surrounding area had been developed for residential purposes and the local population had increased substantially. They noted that Beijing's biggest government housing project (*liangxianfang*) was being constructed 1.5 km from the incinerator, further increasing the local population. Much to the chagrin of local residents, the EIA had stated that the surrounding area was unsuitable for the construction of luxury dwellings, but apparently suitable for building housing for low- and middle-income families. Gaoantun campaigners also questioned details in the EIA, including the distance of residences to the incinerator site, and technical matters relating to emissions generated from waste incineration. The administrative review application, therefore, also demonstrated a good understanding of technical and legal issues related to waste incineration, with campaigners benefiting from Xia Jun's experience from the Liulitun case. Campaigners also claimed that public participation was lacking and that SEPA had 'not only harmed our environmental rights and interests, but had also violated national laws' (Zhou *et al.*, 2007). According to local residents, SEPA had solicited public opinion for the incinerator from 50 nearby rural dwellers who were unlikely to oppose the incinerator. Although this technically met the loose public participation requirements under the EIA Law, anti-incinerator campaigners felt aggrieved that they had not been consulted. In January 2008, SEPA released its administrative review decision in which it upheld the legality of the 2004 EIA. Unsatisfied, residents appealed to the State Council. They accused SEPA of violating the Administrative Review Law by failing to provide sufficient evidence to support its decision. Residents demanded to view the full EIA report, but the State Council supported SEPA's position that the full report contained company secrets and should therefore not be disclosed.

As well as utilizing legal channels, Gaoantun residents organized several protest 'strolls' in the locality.[10] On 30 August 2008, local residents blocked a rubbish transportation vehicle in Gaoantun. Although residents used the protest to highlight opposition to the incinerator, it was mainly sparked by the stench from the landfill, which was processing additional waste from the Olympics.[11] This protest, deliberately staged after and not during the Olympics, won concessions for the residents regarding the stench.[12] Several days later, the Chaoyang District Government formally apologized and promised to invest 91 million Renminbi to resolve the problem within 20 days. Officials also met residents to discuss the incinerator. However, according to one campaigner, officials treated them with disdain, and shocked residents by apparently not having heard of dioxins (Interview with Gaoantun campaigner, Beijing, 28 November 2010).

Gaoantun residents ultimately failed to prevent the incinerator from operating. There were several important differences with the Liulitun case. For example, whereas the Municipal EPB had handled the Liulitun EIA, the Gaoantun EIA had been approved by SEPA. In requesting SEPA to overturn its own decision rather than the decision of a subordinate unit, Gaoantun residents were unable to obtain the kind of support enjoyed by their Liulitun counterparts.[13] Furthermore, the historical disagreement between environmental protection departments and local government that contributed to SEPA's opposition to the Liulitun incinerator was not evident in the Gaoantun case. Third, and most importantly was that the incinerator had already been built. To quote a Chinese proverb used by a Liulitun campaigner in relation to the Gaoantun case, 'the rice has been cooked' (Interview with Liulitun resident, 6 June 2011). This pushed the issue into a different level politically and limited external support. For instance, it would have been far more difficult for a CPPCC representative or government department to call for the project's suspension because that would lead to huge financial losses. In the words of one campaigner, 'no leader would dare [order the incinerator's demolition] . . .' (Interview with Gaoantun campaigner, Beijing, 28 November 2010). The feeling that the government would never reverse its decision arguably undermined the campaign, as many residents became resigned to the incinerator's existence and eventually gave up. This low morale was partly reflected in the nature of the campaign, with several interviewees describing Gaoantun residents as being less united and less persistent than those at Liulitun (Interview with Liulitun residents, 6 June 2011; interview with waste expert, 7 June 2011). One campaigner complained that neighbours were willing to 'free ride' on the efforts of several dedicated individuals (Interview with Gaoantun campaigner, Beijing, 28 November 2010). Residents were also more inclined to pursue their own individual cases. For example, in late 2008 local resident Zhao Lei filed a lawsuit against the waste operators after suffering from respiratory problems that she linked to the waste site. Zhao lost the case and subsequently sold her property and moved away. In addition, although Liulitun residents also came under pressure from the state, this was more pronounced at Gaoantun, with plainclothes police warning HDAG organizers that they would 'bear the consequences of their actions' (Interview with Gaoantun campaigner, 6 June 2011). Finally, Gaoantun residents predominantly consisted of young white-collar workers who lacked time to spend on research and campaigning, and who did not benefit from the same political connections enjoyed by some Liulitun residents.

Asuwei: 'Towards a Win–Win Situation?'[14]

Asuwei, in northern Beijing's Changping District, has hosted the city's largest landfill since it was constructed in 1994. It was subsequently chosen as a waste incineration site as part of the Beijing DRC Domestic Waste Plan. Residents from nearby Aobei, a wealthy neighbourhood of expensive villas, learned about the incinerator in 2009 when one of them visited the township government office to complain about traffic noise. By chance this resident saw a notice (*gongshi*) inviting public comments on the proposed incinerator. Denied permission to photocopy the notice, one of only three that had been posted in the entire city, she photographed it and distributed it amongst her neighbours.

Initially, Aobei residents framed their opposition to the incinerator as a 'rights upholding' campaign in which, similar to the other two cases, they challenged the siting decision from a legal perspective. Local residents established a 'rights upholding small group' and created two websites for exchanging information. From early August 2009, residents utilized *shangfang* and *xinfang* petitioning channels. Campaigners were unhappy with the lack of public participation in the initial siting decision, and argued that Asuwei was an unsuitable site due to its being upstream and upwind of densely populated areas. After failed attempts to meet local officials, several residents drove their cars around the area displaying anti-incineration banners. In response, officials from the municipal, district, and township governments met with residents and defended the legality of the EIA. After the meeting, residents commented that officials had been unable to answer their questions (*Southern Metropolis*, 2010). Despite having defended the original EIA report, on 14 August officials placed a notice in the *Beijing Daily* newspaper inviting public comment on the incinerator. After residents expressed their opinions through this channel, the BMCCAE suspended construction until a new EIA was approved. Despite this concession, on 4 September over 100 Aobei residents in 60 cars converged on the Agricultural Exhibition Centre where the Asuwei Circular Economy Park was being honoured at the 2009 Beijing Environment and Hygiene Fair, and erected anti-incineration banners. A few residents were arrested and imprisoned for several days. Meanwhile, the BMCCAE opened a temporary office in the local township government to collect residents' views in the hope of diffusing the situation.[15]

After the protest, residents began questioning their adversarial 'rights upholding' approach that challenged the incinerator siting decision. In part, they were dissuaded by the authorities' strong response to their public protest. At the same time, a change in approach was necessitated by problems in mobilizing sufficient support (Interview with Asuwei campaigner, 29 November 2010). Compared with densely populated Liulitun and Gaoantun, which had many residents prepared to take to the streets and pressurizing the authorities, Aobei residents were much fewer in number. Many Aobei villas were second homes, and therefore were sometimes unoccupied. Furthermore, wealthy Aobei residents failed to win support of nearby communities. This was partly because nearby villagers were happy to accept compensation from the government. Some even carried out home improvements in the hope that they would subsequently obtain more compensation if and when the government relocated them. It was even suggested that poorer communities enjoyed seeing their rich Aobei neighbours suffering.

In response to these difficulties, Aobei residents changed their approach and began emphasizing their desire to work *with* the government in finding a solution to the waste problem. Aobei residents portrayed themselves as helping the government find a solution to the MSW issue from a public interest perspective rather than from a NIMBY perspective. By building up their own expertise and downplaying the 'rights upholding' position, campaigners hoped to become participants in the policy process rather than opponents of the government (Interview with Asuwei campaigner, 29 November 2010). Similar to Liulitun (and, to a lesser extent, Gaoantun) residents, Aobei residents built up technical expertise concerning MSW management. Residents formed a 'research small group' that produced a very detailed 64-page report into waste incineration (Beijing Aobei Volunteers' Research Group, 2009). This benefited from the fact that one resident had strong professional knowledge of the waste incineration industry. In contrast with the Liulitun 'Opinion Letter', which predominantly focused on the attacking the government's 'irresponsible' siting decision, Aobei residents' report only criticized the Asuwei incinerator siting decision very briefly. The main part of the report compared China's waste incineration situation with other countries in order to highlight more generic problems associated with the former's waste incineration policy. Problems faced by China included that:

- Waste incinerator emission standards are too low; for example, the report claimed that EU dioxin emission standards are 10 times stricter than China.
- The legal framework for waste incineration is poorly developed, weak and ineffective.
- The scale of incinerators being built in China is much greater compared with overseas ones.

- There are major problems with the Chinese incinerator technology and management, particularly the lack of incinerator supervision.

The report concluded that China's 'national circumstances' (*guoqing*) were unsuitable for waste incineration.

Rather than merely criticize Chinese waste incineration policy, the report made a number of suggestions on improving waste management. One suggestion involved transporting waste by train to Inner Mongolia and burying it in the desert. They also suggested that the government focus on waste reduction, and learn from the experiences of Canberra, Edmonton, San Francisco, and the Brazilian Business Commitment for Recycling. In addition, the report stated that, 'we hope the government can be a bit more transparent, and give us truthful information' (*Business Watch*, 2010). In that way, citizens would be able to check and balance corrupt and dishonest officials whom they suspected would cut corners to keep costs down, including by failing to burn waste at a high enough temperature. According to this view, residents were not opposed to waste incineration as long as the process was well managed, which meant having advanced technology, strong enterprise management, government and public supervision, and transparency. According to one interviewee, if these conditions were met then local residents would not oppose the incinerator (Interview with Asuwei campaigner, Beijing, 29 November 2010).

The government was reportedly surprised by the depth and quality of the campaigners' report, which residents have updated regularly (Interview with Asuwei campaigner, Beijing, 29 November 2010). Interaction between campaigners and officials was further facilitated by the involvement of waste expert Wang Weiping, who met Aobei residents on several occasions and who became a go-between for residents and the government. At Wang's suggestion, the Beijing Municipal Government invited prominent campaigner Huang Xiaoshan on a 10-day visit to Japan and Macau to inspect waste treatment facilities. The state media portrayed this as a highly important and symbolic gesture. One media report suggested that the government's decision to invite Huang shows that the government and public have found a common language and can cooperate together towards a 'win–win' situation (*The Beijing News*, 2010a). However, suspicions were also voiced that Huang was being co-opted by the government (*Southern Metropolis*, 2010). One Liulitun campaigner stated that it is easier for people such as Huang to make connections with the government because of their high social status and personal connection (*guanxi*) networks (Interview with Liulitun campaigner, Beijing, 5 June 2011). They could also use these connections to fight against the incinerator whilst publicly portraying themselves as promoters of the public interest (Interview with waste expert, 3 July 2012).

In April 2010 municipal officials launched the 'Be a Civilised and Polite Beijing Person—Carry Out Waste Sorting and Reduction' campaign, which included setting up waste sorting pilot projects in 600 communities. Asuwei residents credit their own efforts with this development (Interview with Asuwei campaigner, Beijing, 29 November 2010), although it is difficult to determine the extent of their influence. In any case, through their research and willingness to look at the waste problem from a broader public interest perspective, Asuwei residents showed that the public need not only be an impediment to the government's waste policy. In April 2011, it was announced that the construction of the Asuwei incinerator would begin that year. Yet as of June 2012 construction had not begun, and interviewees were unsure as to what would happen (Interview with waste expert, 3 July 2012; interview with NGO staff member, 3 July 2012). It remains unclear whether the incinerator will be built and, if so, whether it will meet the high standards requested by residents.

Discussion

Since the Liulitun anti-incineration campaign began in 2006, Beijing residents and officials have undergone a learning process. Anti-incineration campaigners have learnt from each other and in some cases have become lay experts. Officials have learnt lessons and have begun to modify their incineration strategy.

Citizens as Experts

Tension between government and industry experts and citizens regarding who should make decisions concerning potentially polluting projects has been a key feature of contention over waste facility siting in Western countries (Fischer, 2000). Fischer (2000) has argued that expertise has become a key dividing line in struggles over such projects. Numerous cases have been documented of 'citizen experts' (Tesh, 1999) developing their own expertise in technical and legal aspects of siting issues to counter perceived bias from government and industry experts (see, for example, Fletcher, 2003; Freudenberg, 1984; Shemtov, 1999).

One of the most striking aspects of the three Beijing campaigns was how citizens developed legal and technical expertise through self-study and contact with experts. This 'expert strategy' was particularly prominent in the Liulitun and Asuwei campaigns, and was adopted for several reasons. First, campaigners required to show that they understood the issues. One Beijing Municipal People's Congress (BMPC) member and incineration proponent claimed that the public 'do not understand' technical and scientific information, therefore implying that public disclosure on emissions information was unnecessary (Interview with BMPC member, Beijing, 10 June 2011). Confronted by this type of attitude, campaigners felt that to appear credible and make the government (and wider public) listen, they had to argue their case based on 'adequate reasoning and evidence' (*you li you ju*) (Interview with Liulitun campaigner, Beijing, 5 June 2011). Second, campaigners' expert strategy was a response to the discovery that many local officials were poorly informed and lacked technical and policy expertise concerning incineration. In other words, citizens accumulated local knowledge partly to prove that siting decisions had not been predicated upon sound technical knowledge (or even basic common sense), but had been taken arbitrarily. One interviewee described this lack of knowledge as the government's 'Achilles' heel', and suggested that, '"officials" [waste incineration] knowledge is behind that of the public' (Interview with waste expert, Beijing, 7 June 2011). Through studying legislation and accumulating local knowledge, participants in all three campaigns highlighted basic errors and illegalities regarding site selections. Particularly in the Asuwei case, this extended to a broader critique of the MSW policy. One campaigner claimed that the government is 'lazy' and that officials simply search for the easiest solution, namely burning rubbish without taking an holistic view of the issue (Interview with Asuwei campaigner, Beijing, 29 November 2010). This development of counter-expertise made it harder for officials to marginalize residents as irrational NIMBY campaigners. As one campaigner stated, 'I'm not just going to say 'you shouldn't build it here'; I'm going to tell you *why* you shouldn't build it here' (Interview with Asuwei campaigner, Beijing, 29 November 2010). Third, the expert strategy was utilized to benefit from divisions within the scientific community. Campaigners in all three cases approached experts they knew to be sceptical about, or downright opposed to, incineration so as to obtain technical information and support. In an interesting juxtaposition between *shangfang* and *xinfang*, Liulitun campaigners referred to this as one strand of a '*baifang*' ('visit') strategy. Some experts such as Zhao Zhangyuan spoke out publicly against incineration and furnished campaigners with useful technical information. Expert support was particularly important given the relatively weak capacity of Chinese environmental NGOs in this respect (Interview with NGO staff, Beijing, 8 June 2011). A final rationale for pursuing an expert strategy was to depoliticize the issue by focusing on legal and technical matters. By performing the role of 'citizen experts' offering constructive suggestions concerning waste policy, residents could portray themselves as concerned citizens supportive of the public interest. Campaign participants distinguished between correcting perceived injustices through legitimate 'rights upholding' and political activism. A Gaoantun campaigner stated that, 'the overall principle was that we don't support people sacrificing themselves; violent confrontation with the government can't solve any problems, it will only intensify the contradictions (*maodun*), people's personal safety would be jeopardised' (Interview with Gaoantun campaigner, Beijing, 28 November 2010). Indeed, the second strand of Liulitun campaigners' *baifang* strategy involved visiting government officials in an 'informal' rather than overt 'petitioning' *shangfang* or *xinfang* context in order to discuss incinerator-related issues. This strategy was utilized more fully by Asuwei residents, culminating in Huang Xiaoshan's inclusion in the Japan study tour.

Campaigners pursued an expert strategy to different degrees. For example, unlike Liulitun and Asuwei residents, Gaoantun campaigners did not publish research reports. According to one prominent Gaoantun campaigner, Liulitun residents included many retired intellectuals, which enabled them to conduct research and built up counter-expertise against the state (Interview with Gaoantun campaigner, Beijing, 6 June 2011). Asuwei residents also included individuals with time and research skills. Participants in the Liulitun and Gaoantun cases focused more on accumulating legal expertise in order to challenge the siting decisions, whilst supplementing this approach with other tactics including public protest. One interviewee said that 'if [citizens'] tactics are too soft, the government will simply ignore you' (Interview with Liulitun campaigner, Beijing, 29 July 2009). Although Aobei campaigners initially opposed the Asuwei incinerator siting from an adversarial 'rights upholding' approach, they subsequently decided that a policy advocacy approach was more effective for their circumstances. This has arguably helped to stimulate wider discussion about waste treatment in Beijing; however, it may not be enough to derail the Asuwei incinerator.

Government Learning

Government officials in Beijing have adapted their tactics with regards to incinerator siting in response to societal opposition. Faced by citizens who are becoming increasingly vigilant concerning waste incineration, some local officials have tried to show sensitivity to citizens' concerns. For example, Chen Yong has publicly stressed the need for governments to publicly disclose information concerning incinerator emissions, whilst also calling for improvements in public participation channels (*The Beijing News*, 2010b). The Gaoantun incinerator, which remains the only operational incinerator in Beijing, is open to the public every Thursday. Furthermore, a digital public information screen displaying data about the Gaoantun incinerator's emissions was installed on-site following a suggestion filed by a local BMPC member. In addition, the public has twice been invited to submit comments regarding the Sujiatuo incinerator. Although these developments suggest early steps towards greater transparency and public engagement, in reality they appear superficial, tokenistic, and geared towards public appeasement. For example, on 10 June 2011, I visited the Gaoantun incinerator and observed the public information screen. Although it displayed data about dioxins, the chief health concern for anti-incinerator campaigners in all three cases, this information was based on a sample from 22 February. When questioned about this time lag, the BMPC member accompanying me revealed that measuring dioxins is 'too expensive' (Interview with BMPC member, Beijing, 10 June 2011). Furthermore, information from the screen is not available online, and the screen itself is inside the waste treatment compound and therefore not always accessible to the public. For all intents and purposes, this screen appears to tally more with one campaigner's dismissive remark about it being there to 'look good' (*hao kan*) rather than representing an attempt to improve transparency (Interview with Gaoantun campaigner, Beijing, 28 November 2010).

Similarly, the soliciting of public comments for the Sujiatuo incinerator appears more related to the unique circumstances of that facility than being a serious move towards greater public engagement. Although the Sujiatuo incinerator will replace the one planned at Liulitun, the latter site will play a role in separating waste before it is sent to Sujiatuo. Given the strong public opposition to the Liulitun incinerator, Haidian District Government officials are keen to minimize any further potential opposition by following proper procedure regarding public opinion solicitation (Interview with NGO member, Beijing, 8 June 2011). At the same time, evidence suggests that local officials in Beijing and elsewhere are learning to circumvent opposition by selecting sites in more remote locations where opposition is weak, and by further limiting transparency. In 2011, a BMCCAE official stated that, 'if you want to be successful in selecting a location [for an incinerator], before construction begins you can't reveal it. This is because as soon as you choose a site, nearby residents will oppose it' (*The Beijing News*, 2011). Rather than embrace transparency and public consultation, in 2010 municipal government officials pushed through the approval of an incinerator, set to be Asia's largest, at Lujiashan on a site formerly owned by the Capital Steel Company in western Beijing just outside the Sixth Ring Road. The Lujiashan incinerator was approved extremely quickly even by Chinese standards (*Southern Weekend*, 2011). Zhao Zhangyuan claimed that it obtained the roughly 80 necessary approval stamps within just 3 months, during which time public comment was not solicited (*Southern Weekend*, 2011). The full title of the project, namely 'Beijing Capital Steel Biomass Power Project', makes no reference to incineration. The Sujiatuo incinerator's official title of 'Beijing City Haidian District Circular Economy Industrial Park Renewable Energy Electricity Generation Plant Project' is similarly reminiscent of 'incinerators in disguise' found in other countries. The Lujiashan incinerator was unveiled on 23 October 2010 by senior officials in a public show of support. The strategy of rushing through incinerators to circumvent opposition may lead to lower standards due to a less rigorous planning process. Finally, there are indications that officials in Beijing are pursuing a strategy of expanding existing incinerators rather than attempting to site facilities in new locations. For example, in 2011 approval was granted for a 'second phase' (*erqi*) of the Gaoantun incinerator to be constructed, with a daily capacity of 1800 ton. A third phase is apparently also under consideration (Interview with NGO staff member, 3 July 2012). A second phase of the Lujiashan incinerator is also being planned, which would increase the capacity to at least 5000 ton per day. Although this helps officials to limit opposition, it also raises the stakes for communities reluctant to allow incinerators to gain a foothold in their neighbourhoods.

Conclusion: Implications for Waste Incineration Policy in China

The three case studies examined in this article highlight the limitations of a non-participatory MSW policy in a context where ordinary citizens are increasingly aware of their legal 'right' to be consulted and informed about siting decisions. Top-down, opaque decision-making processes with regards to the siting of unpopular waste incinerators have collided with citizens who mobilize strongly in defence of their interests using the rhetoric of 'rights-upholding', and who develop their own technical counter-expertise against the state. Due to societal opposition, Beijing will fail to achieve its 11th FYP goals for MSW management. In July 2010, Chen Yong admitted that the 4:3:3 objective would not be achieved until 2015, 4 years behind schedule (Chen, 2010). The short-term target was downgraded, with officials aiming to achieve a ratio between incineration, composting, and landfill of 2:3:5 by 2012 (Chen, 2010). Chen identified disagreements, problems in site selection, and a lack of public understanding as key factors that had inhibited Beijing's MSW policy (*The Beijing News*, 2010b). From the government's perspective, the public has, therefore, undermined its top-down MSW strategy in the short term.

Although Chinese anti-incineration campaigns and government responses display their own unique characteristics, there are also striking similarities with earlier anti-siting campaigns in the West. However, in the longer term it is highly unlikely that public opposition to incinerator siting decisions will undermine the promotion of incineration in China to the same extent as incineration was undermined in, for example the USA. Despite opposition, officials within the Beijing municipal and district governments remain committed to incineration. As demonstrated by the Lujiashan, Sujiatuo, and indeed Gaoantun incinerator sitings, government officials in illiberal China are in a very strong position when it comes to pushing through projects in secrecy, and deflecting lawsuits by aggrieved citizens. In addition, anti-incinerator campaigns in China are predominantly localized and NIMBY-focused. Prospects for disparate campaigners to form a movement strong enough to challenge the state's incineration policy seem remote. The CCP remains extremely nervous about geographically disperse groups forming a wider network that could challenge its monopoly on political power (Ho, 2007). That is not to say that campaigners within Beijing and beyond have not begun to network with each other. Internet community bulletin boards, which residents use to share materials and vent opinions, are easily accessible, and can be a useful source of information for campaigners seeking to learn from other campaigns. In addition, the Beijing-based environmental NGO Green Beagle conducts regular waste-related seminars featuring expert speakers and study trips to waste facilities. Campaign leaders from as far afield such as Guangzhou and Shenzhen have participated in these events, along with their Beijing counterparts. They have been able to share their campaigning experiences in person and learn about broader waste management issues. However, this type of networking is low profile and, similar to the campaigns themselves, steers away from any overt political agenda. At least in the short term, the main impact of this informal networking is to help middle-class homeowners resist incinerator sitings in their own backyards. Although beneficial to middle-class homeowners, this may have implications for environmental equality in China as local officials adapt their incinerator siting tactics. Part of the Beijing municipal government's rationale for siting incinerators at the more sparsely populated Sujiatuo and Lujiashan is the fact that many local residents are relatively poor peasants. These people are more easily placated through compensation and are therefore less likely to oppose an incinerator (Interview with NGO staff member, 8 June 2011). The decision to build Beijing's largest *liangxianfang* next to the Gaoantun incinerator also suggests that less resourceful poorer residents may have to bear the brunt of China's incineration expansion. The locating of incinerators in areas of social deprivation has been a key driver of the environmental justice movement in the USA (Rootes, 2009). Ironically, largely middle-class anti-incineration campaigns of the type examined by this article that demand a 'fairer' decision-making process could lead to greater environmental injustice in China, perpetuating the notion that MSW, 'tumbles downhill to settle in places of least resistance, among the poor and disenfranchised' (Royte 2005, p. 40). Given China's growing waste crisis and public opposition to incineration, it is likely that contention will increasingly be a factor in its MSW policy.

Acknowledgements

The author would like to thank the following people for their comments and suggestions: Genia Kostka, Arthur Mol, Rachel Stern, colleagues from the City University of Hong Kong Governance in Asia Research Centre, and the anonymous reviewers. Any errors remain the author's alone. The work described in this paper was fully supported by a grant from the College of Liberal Arts and Social Sciences, City University of Hong Kong (Project No. 9610162).

Notes

1. The 2% incineration figure apparently refers to a small-scale, informal incineration. One government-affiliated expert expressed doubt about the reliability of these statistics (Interview with waste expert, Beijing 7 June 2011).
2. The 'city area' generates over two-thirds of Beijing's MSW.
3. According to *The Guardian*, less than 4% of Beijing's waste is recycled (*The Guardian*, 2010).
4. Although the decision to build an incinerator was made in 2005, residents reportedly only became aware of it in November 2006 after they saw the Beijing Municipality 11th FYP (Beijing Haidian District Residents and Enterprise Staff, 2009).
5. In using the term 'unscientific', campaigners invoked the language of President Hu Jintao and his calls for 'Scientific Development'.

6. Authorities are only required to release abridged versions of EIA reports to the public.
7. According to one resident, they converged on SEPA headquarters not to protest but to 'celebrate World Environment Day' (Interview with Liulitun resident, 4 July 2012). However, the intention was undoubtedly to place pressure on the government.
8. This is the conclusion reached by a meeting of anti-incineration experts and NGO representatives (meeting minutes on file with author).
9. No evidence was provided to support this claim.
10. Without prior approval from the authorities, protests are technically illegal. Urban residents try to get around this by dubbing their actions 'strolls'.
11. According to one Liulitun resident, Gaoantun had to deal with much more waste during the Olympics, as the government was reluctant to send extra waste to Liulitun for fear of provoking unrest (Interview with Liulitun campaigner, 6 June 2011).
12. In contrast, Fletcher (2003) has documented how activists in Toronto used the city's bid to host the 2008 Olympics to apply pressure to the city government.
13. Furthermore, Pan Yue was reportedly sidelined during 2008, apparently because of his outspokenness and willingness to go after large-scale projects that had violated EIA procedures.
14. This account draws heavily from *Business Watch* (2010).
15. According to one waste expert, this was a sign that government officials had learned from the Liulitun and Gaoantun cases by making themselves (at least appear to be) more accessible to local residents (Interview with waste expert, Beijing, 7 June 2011).

References

Asian Development Bank (2009, September 3) *ADB supports clean waste-to-energy project in the PRC.* News Release. Available at http://www.adb.org/media/Articles/2009/12974-chinese-clean-energies-projects/ (accessed 12 November 2009).

Atlas, M. (2001a) Testing for environmental racism again: An empirical analysis of hazardous waste management capacity expansion decisions. Paper presented at the Annual Meeting of the American Political Science Association, San Francisco.

Atlas, M. (2001b) Safe and sorry: Risk, environmental equity, and hazardous waste management facilities, *Risk Analysis*, 21(5), pp. 939–954.

Beijing Aobei Volunteers' Research Group (2009) *Zhongguo chengshi huanjing de shengsi jueze* [The life and death decision for China's cities' environment]. Available at http://www.eiafans.com/thread-57712-1-1.html (accessed 24 November 2010).

Beijing DRC (2007, April 10) *Beijing Shi 'Shiyiwu' shiqi shenghuo laji chuli sheshi jianshe guihua shishi fang'an* [Beijing Municipality solid waste treatment plan for the 11th Five-Year Plan period]. Available at http://www.bjpc.gov.cn/fzgh/news/200704/t167502.htm (accessed 11 December 2010).

Beijing Haidian District Residents and Enterprise Staff (2009, 31 March) *Fandui zai Liulitun jianshe laji fenshao chang yijianshu* [Opinion letter for opposing the construction of a waste incinerator in Liulitun]. Available at http://wenku.baidu.com/view/b388b93b87c24028915fc30a.html (accessed 14 February 2010).

Business Watch (2010, April 8) *Boyi Asuwei* [The Asuwei game]. Available at http://finance.sina.com.cn/leadership/mroll/20100408/14337711690.shtml (accessed 12 December 2010).

Chen, Y. (2010, July 30) *Guanyu shenghuo laji chuli gongzuo jinzhan qingkuang de baogao* [Report concerning work progress in relation to the domestic waste treatment situation]. Report delivered at the 19th meeting of the Standing Committee for the Beijing Municipality 13th People's Congress, Beijing. Available at http://www.bjmac.gov.cn/pub/guanwei/K/K1/201007/t20100730_19234.html (accessed 14 December 2010).

China Daily (2009, June 16) *China's urban population exceeds 600 million.* Available at http://www.chinadaily.com.cn/china/2009-06/16/content_8288412.htm (accessed 12 November 2009).

Cole, L. & Foster, S. (2001) *From the Ground Up: Environmental Racism and the Rise of the Environmental Justice Movement* (New York: New York University Press).

Davies, A. R. (2008) *The Geographies of Garbage Governance* (Aldershot: Ashgate).

Eurostat (2010) Waste statistics. Available at http://epp.eurostat.ec.europa.eu.statistics_explained/index.php/Waste_statistics (accessed 21 June 2011).

Feng, Y. (2009) *Bian Zuo Huanbao Bian Sahuang: Xie Gei Gongzhong de Huanbao Neican [Protecting the Environment Whilst Lying: An Environmental Protection Internal Reference Written for the Public]* (Beijing: World Knowledge Press).

Feng, Y. (2010) Laji chuli de gongzhong canyu: chengxu zhengyi he jishu zhengyi [Public participation in waste treatment: procedural justice and technological justice], in: D. Yang (Ed) *Huanjing Lü Pi Shu: Zhongguo Huanjing Fazhan Baogao (2010) (Annual Report on Environment Development of China (2010))*, pp. 190–198 (Beijing: Social Sciences Academic Press).

Fischer, F. (2000) *Citizens, Experts and the Environment: The Politics of Local Knowledge* (Durham, NC: Duke University Press).

Fletcher, T. (2003) *From Love Canal to Environmental Justice: The Politics of Hazardous Waste on the Canada–U.S. Border* (Peterborough, Ontario: Broadview Press).

Freudenberg, N. (1984) Citizen action for environmental health: Report on a survey of community organizations, *American Journal of Public Health*, 74(5), pp. 444–448.

Freudenberg, N. & Steinsapir, C. (1992) Not in our backyards: The grassroots environmental movement, in: R. Dunlap & A. Mertig (Eds), *American Environmentalism: The US Environmental Movement, 1970–1990*, pp. 27–37. (Philadelphia, PA: Taylor & Francis).

Gottlieb, R. (1993) *Forcing the Spring: The Transformation of the American Environmental Movement* (Washington, DC: Island Press).

Hess, D. (2007) *Alternative Pathways in Science and Industry: Activism, Innovation, and the Environment in an Era of Globalization* (Cambridge, MA: MIT Press).

Ho, P. (2007) Embedded activism and political change in a semiauthoritarian context, *China Information*, 21(2), pp. 187–209.

Home Defence Action Group (2008, August 26) *Gaoantun shenghuo laji fenshao chang xingzheng fuyi zhi lu* [The road to administrative review for the Gaoantun domestic waste incinerator]. Available at: http://yuedulaiyin.soufun.com/bbs/1010036393~-1~5701/77163886_77163886_%7Bpage%7D.htm (accessed 7 December 2010).

Innes, J. & Booher, D. (2004) Reframing public participation: Strategies for the 21st century, *Planning Theory & Practice*, 5(4), pp. 419–436.

Johnson, T. (2010) Environmentalism and NIMBYism in China: Promoting a rules-based approach to public participation, *Environmental Politics*, 19(3), pp. 430–448.

Joppke, C. (1993) *Mobilizing Against Nuclear Energy: A Comparison of Germany and the United States* (Berkeley, CA: University of California Press).

Liu, J. (2007, June 22) Small, yet brave, *China Dialogue*. Available at http://www.chinadialogue.net/article/show/single/en/1117-Small-yet-brave (accessed 1 July 2007).

McAvoy, G. (1998) Partisan probing and democratic decision making: Rethinking the NIMBY syndrome, *Policy Studies Journal*, 26(2), pp. 274–292.

McAvoy, G. (1999) *Controlling Technocracy: Citizen Rationality and the NIMBY Syndrome* (Georgetown: Georgetown University Press).

New Century (2012, 10 January) *"Shi Er Wu" laji fenshao Da Yue Jin huang shiqi* ["12th Five-Year Plan" waste incineration Great Leap Forward golden age]. Available at http://news.solidwaste.com.cn/view/id_37368 (accessed 31 July 2012).

Nissim, I., Shohat, T. & Inbar, Y. (2005) From dumping to sanitary landfills: Solid waste management in Israel, *Waste Management*, 25(3), pp. 323–327.

OECD (2007) *OECD Environmental Performance Reviews: China* (Paris: OECD).

People's Daily (2009, April 1) *Laji weicheng* [Waste surrounds the cities]. Available at http://society.people.com.cn/GB/97741/125117/9065450.html (accessed 3 April 2011).

Perry, E. & Selden, M. (2003) Introduction: Reform and resistance in contemporary China, in: E. Perry & M. Selden (Eds), *Chinese Society*, 2nd ed., pp. 1–22 (London: RoutledgeCurzon).

Petts, J. (2001) Evaluating the effectiveness of deliberative processes: Waste management case-studies, *Journal of Environmental Planning and Management*, 44(2), pp. 207–226.

Rootes, C. (2009) Environmental movements, waste and waste infrastructure: An introduction, *Environmental Politics*, 18(6), pp. 817–834.

Rootes, C. & Leonard, L. (2009) Environmental movements and campaigns against waste infrastructure in the United States, *Environmental Politics*, 18(6), pp. 835–850.

Rootes, C. & Leonard, L. (Eds) (2010) *Environmental Movements and Waste Infrastructure* (London: Routledge).

Royte, E. (2005) *Garbage Land: On the Secret Trail of Trash* (New York: Back Bay Books).

SEPA (2007a, 7 June) *Huanbao Zongju: Liulitun laji fadian xiangmu yingyu zanjian. Jiang dui Xiamen quan quyu jinxing guihua huanping* [SEPA: Liulitun waste to electricity generation project should be suspended. A strategic EIA for the entire Xiamen district will be carried out]. Available at http://www.sepa.gov.cn/xcjy/zwhb/200706/t20070607_104698.htm (accessed 14 June 2007).

SEPA (2007b, 12 June) *Qian Zuosheng he Li Huilan deng 6 ren de xingzheng fuyi jueding shu* [Administrative review decision for the 6 Applicants including Qian Zuosheng and Li Huilan]. Available at http://www.sepa.gov.cn/law/fzjs/xzfy/200706/t20070612_105077.htm (accessed 14 June 2007).

SEPA & NDRC (2006) *Guanyu jin yi bu jiaqiang shengwuzhi fadian xiangmu huanjing yingxiang pingjia guanli gongzuo de tongzhi* [Notice regarding strengthening of EIA management work for projects involving electricity generation from biomass]. Available at http://www.mep.gov.cn/info/bgw/bwj/200809/t20080908_128308.htm (accessed 2 February 2012).

Shemtov, R. (1999) Taking ownership of environmental problems: How local NIMBY groups expand their goals, *Mobilization: An International Journal*, 4(1), pp. 91–106.

Shevory, T. (2007) *Toxic Burn: The Grassroots Struggle Against the WTI Incinerator* (Minneapolis, MN: University of Minnesota Press).

Southern Metropolis (2010, 5 March) *Beijing Aobei bieshu qu yezhu PK Yazhou zui da laji fenshao fadian chang* [Beijing Aobei villa homeowners PK Asia's biggest waste incinerator]. Available at http://solidwaste.chinaep-tech.com/special-topic/74731.htm (accessed 21 June 2011).

Southern Weekend (2011, February 23) *Laji fenshao, yuanhe "mingxiu zhandao, andu chencang." "Liulitun" zhong qijian, "Lujiashan" ji shangma* [Why is waste incineration akin to "mending the road in the light, entering into secret liaison in the dark?" "Liulitun" is finally abandoned, "Lujiashan" in a rush to begin]. Available at http://www.infzm.com/content/55573 (accessed 3 April 2011).

Tesh, S. (1999) Citizen experts in environmental risk, *Policy Sciences*, 32, pp. 39–58.

The Beijing News (2010a, February 21) *Beijing dizhi jian laji fenshao chang wangyou huo zhengfu yaoqing fu Ri kaocha* [Beijing netizen opposed to waste incinerator receives government invitation to visit Japan for investigation]. Available at http://society.people.com.cn/GB/41158/10989380.html (accessed 25 November 2010).

The Beijing News (2010b, July 31) *Beijing san da lajichang wufa anqi wangong xiangguan lingdao yaoqiu jiakuai jianshe* [Work on three big waste facilities in Beijing unable to be completed on time, relevant leaders request speeding up of construction]. Available at http://news.dichan.sina.com.cn/bj/2010/07/31/192730.html (accessed 10 December 2010).

The Beijing News (2010c, 1 November) *Zhongguo duo chengshi laji weicheng, guanli quexian zhi luan paifang* [Multiple Chinese cities are surrounded by waste, management inadequacy results in dumping]. Available at http://ep.chinaluxus.com/Erv/20101101/17969.html (accessed 1 December 2010).

The Beijing News (2011) *Asuwei laji xunhuan jingji yuan jinnian jiang kaigong* [Construction of the Asuwei waste circular economy park will begin this year]. Available at http://www.bjnews.com.cn/news/2011/04/29/120931.html (accessed 2 August 2012).

The Guardian (2010) Beijing to sweeten stench of rubbish crisis with giant deodorant guns. Available at http://www.guardian.co.uk/environment/2010/mar/26/beijing-rubbish-deodorant (accessed 21 June 2011).

The New York Times (2005, 12 May) How do Japanese dump trash? Let us count the myriad ways. Available at http://www.freerepublic.com/focus/f-news/1401419/posts (accessed 21 June 2011).

The New York Times (2008, 4 July) Waste-to-energy plants a waste of energy, recycling advocates say. Available at http://www.nytimes.com/2008/07/04/business/worldbusiness/04iht-rbogwaste.html (accessed 21 June 2011).

Walsh, E., Warland, R. & Smith, D. (1997) *Don't Burn It Here: Grassroots Challenges to Trash Incinerators* (University Park, PA: The Pennsylvania State University Press).

World Bank (2005) *Waste Management in China: Issues and Recommendations*, Urban Development Working Papers, East Asia Infrastructure Department, Working Paper No. 9.

Wu, F., Jiang, X. & Yeh, A. G. (2007) *Urban Development in Post-Reform China: State, Market, and Space* (London: Routledge).

Yang, C. (2010) *Laji weiji zou dao shizi lukou* [Waste crisis has reached a crossroads], in: D. Yang (Ed) *Huanjing Lü Pi Shu: Zhongguo Huanjing Fazhan Baogao (2010)* (*Annual Report on Environment Development of China (2010)*), pp. 60–67 (Beijing: Social Sciences Academic Press).

Zhou, Y., Chen, Z., Han, B., Wang, L. & Liu, X. (2007, September 28) *Xingzheng fuyi shenqingshu* [Administrative review application]. Available at http://www.51bolin.com/archiver/?tid-2210.html (accessed 5 December 2010).

Index